MICROWAVE ENGINEERS' HANDBOOK

VOLUME 2

compiled and edited by
THEODORE S. SAAD

co-editors
ROBERT C. HANSEN
GERSHON J. WHEELER

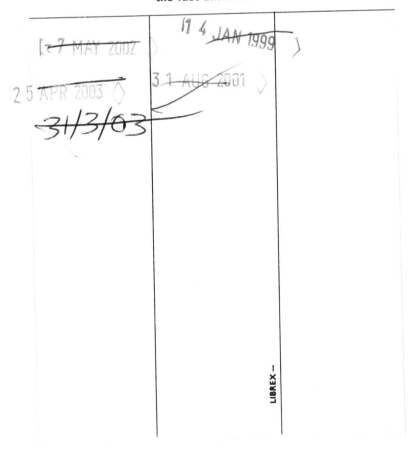

ARTECH HOUSE, INC.
685 Canton Street
Norwood, MA 02062

Printed and bound in the United States of America
by Halliday Lithograph Corporation

Library of Congress catalog card number: 76-168891

MICROWAVE ENGINEERS' HANDBOOK

VOLUME 2

CONTENTS

DIRECTIONAL COUPLERS

MULTI-APERTURE DIRECTIONAL COUPLER DESIGN

1.   Quarter-wave transformer prototype

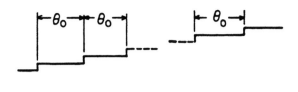

$$\theta_o = \frac{\pi w_q}{4} \qquad \text{where} \qquad w_q = \frac{\omega_2 - \omega_1}{\omega_o}$$

$$s_k = \exp\left[\frac{\rho_k l_n (R)}{\sum_{k=1}^{n} \rho_k}\right] \qquad , k = 1,2,\ldots n$$

Normalized
Impedances: $1 \quad Z_1 \quad Z_3 \text{----} Z_{n-1} \quad Z_n = R$

Junction
VSWR's: $s_1 \quad s_2 \quad s_3 \text{---} s_{n-1} \quad s_n$

Reflection
Coefficients: $\rho_1 \quad \rho_2 \quad \rho_3 \text{----} \rho_{n-1} \quad \rho_n$

$$s_k = \frac{Z_k}{Z_{k-1}} > 1$$

$$S_k = S_{(A-K+1)}$$

$$\rho_k = \frac{s_{k-1}}{s_{k+1}} \qquad \text{For } k = 1, 2,\ldots n$$

where for maximally-flat response (a):-

$$\rho_k = \binom{n-1}{k-1} \qquad \rho_m = \frac{R-1}{2\sqrt{R}} \sin^{n-1}\theta_o \qquad \text{for } \rho_m \ll 1$$

and for Chebyshev equal-ripple response (b):-

$$\rho_k = \frac{n-1}{n-k} \sum_{r=0}^{k-2} \binom{n-k}{r+1} \cdot \binom{k-2}{r} \cos^{2(r+1)}\theta_o$$

$$\rho_m = \frac{R-1}{2\sqrt{R}} \Bigg/ \cosh\left[(n-1)\cosh^{-1}\left(\frac{1}{\sin\theta_o}\right)\right]$$

for $\rho_m \ll 1$

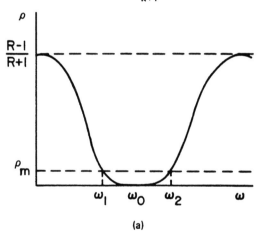

(a)

Alternative derivations:

(i)   Form junction VSWR's by exact synthesis

(ii) Make most central junction VSWR's equal by using a series ρ_k (k=1,2, ... n) formed by the "superimposed array" technique.

2.   Design of Multi-aperture couplers-general

Coupling $= 20 \log\left(\frac{1}{k}\right)$ dB

Choose $R = \left(\frac{1 + \sin\psi}{1 - \sin\psi}\right)^n$

Where $\psi = \frac{2}{n} \sin^{-1} k$

Isolation is given by $20 \log\frac{2}{\rho_m}$ dB - determines value of θ_o.

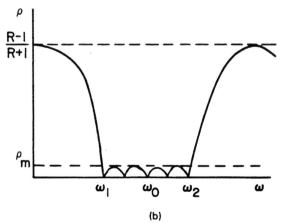

(b)

Courtesy of R. Levy

DIRECTIONAL COUPLERS

For each aperture:

$$K(\omega_c, A) = \frac{\tan\left(\frac{\pi\omega_o}{\omega_o}\right)}{\left(\frac{\pi\omega_o}{\omega_c}\right)} \exp\left[\frac{-2\pi A t}{\lambda_c} \sqrt{1 - \frac{\omega_o^2}{\omega_c^2}}\right]$$

ω_o = Mid band frequency (radians/sec)

ω_c = resonant frequency of the aperture

$\lambda_c = \frac{2\pi v}{\omega}$ resonant wavelength of the aperture

v = = velocity of light

t = common wall thickness

A = empirically determined

For circular holes of diameter d,

$$A \simeq 1 + 0.065\, d/t$$

3. Sidewall couplers

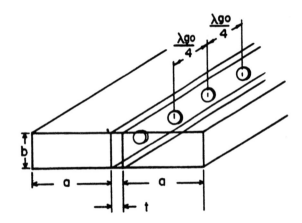

Determine aperture dimensions by solution of the equations:

$$\sqrt{s_k} \cdot \frac{1}{\sqrt{s_k}} = \frac{2\pi\lambda_{go}M_z}{a^3 b} \cdot K\,(\omega_{CH}, A)$$

$s_1, s_2 \cdots s_b$ Prototype junction VSWR's from 1

M_z polarizability of the aperture for the H_z field

λ_{go} mid-band guide wavelength

ω_{CH} ω_c for H_z excitation

For circular holes of diameter d:

$$M_z = \frac{d^3}{6} \qquad\qquad \lambda_{CH} = 1.705\, d$$

4. Broadwall Couplers

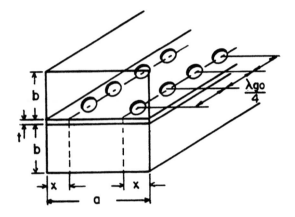

Determine aperture dimensions by solution of the equations:

$$\sqrt{s_k} \cdot \frac{1}{\sqrt{s_k}} = \left(\frac{16\pi M_x \sin^2\frac{\pi x}{a}}{ab\lambda_{go}}\right) + \left(\frac{4\pi\lambda_{go}M_z \cos\frac{2\pi x}{a}}{a^3 b}\right)$$

$$K(\omega_{CH}, A)\; \left(\frac{8\pi\lambda_{go}P_y \sin\frac{2\pi x}{a}}{ab\lambda_o^2}\right) \qquad K(\omega_{CE}, A)$$

$s_1, s_2, \ldots s_n$ prototype junction VSWR's from 1.

λ_o mid-band free-space wavelength

M_x, M_z, P_y polarizabilities of apertures for the H_x, H_z, and E_y fields

ω_{CE}, ω_{CH} resonant frequencies of each aperture for electric and magnetic fields.

For circular holes of diameter d;

$$M_x = M_z = \frac{d^3}{6} \qquad\qquad P_y = \frac{d^3}{12}$$

$$\lambda_{CH} = 1.705d \qquad\qquad \lambda_{CE} = 1.305d$$

Courtesy of R. Levy

2

TEM DIRECTIONAL COUPLERS

1 STEPPED COUPLING SYMMETRIC 90-DEGREE COUPLER

A matched device of theoretically infinite isolation with coupled (C) and transmitted (T) outputs differing in phase by 90 degrees at all frequencies.

Utilizes an odd number (1, 3, 5, etc.) of coupled-line sections and, in theory, is capable of providing any desired bandwidth, mean level (M), and coupling tolerance (δ), by using a sufficient number of sections. Bandwidths up to approximately 10:1 can be realized in practical structures with moderate mean coupling values, with larger bandwidths being difficult to realize due to the high coupling required in the center sections. An exact synthesis procedure and comprehensive element value tables are available (1), (2).

asymmetric coupler provides an optimum design from the viewpoint of providing maximum bandwidth for a given mean level (M), coupling tolerance (δ), and number of sections (n). Practical bandwidth limitations are similar to the symmetric coupler, however, greater bandwidth can be obtained for a given maximum coupling coefficient than with a symmetric design. Detailed exact design and synthesis procedures as well as element value tables are given in (3) and (4). Approximate 90-degree or 180-degree phase difference between coupled (C) and transmitted (T) ports can be obtained by use of an all-pass phase equalizer (5).

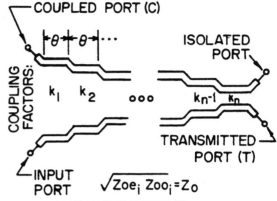

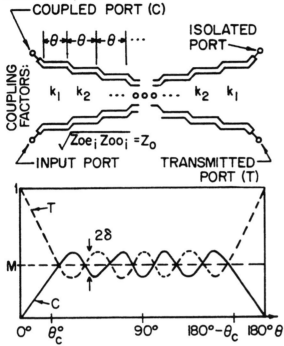

3 RE-ENTRANT COUPLED SECTION (6)

A novel configuration for achieving a high coupling coefficient in a practical structure. Useful as the center section in a multi-section symmetric coupler or as the end section in an asymmetric coupler. Can also be constructed in multi-layer stripline (7). A detailed description of the structure and practical applications can be found in (6).

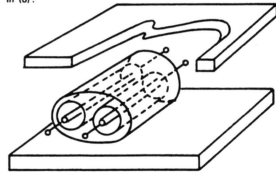

2 STEPPED COUPLING ASYMMETRIC COUPLERS

Coupling and isolation characteristics are similar to that of the symmetric structure, however, no constant phase difference exists between the coupled (C) and transmitted (T) ports. Can be constructed using any number of coupled sections (n), and, in theory, is capable of providing any desired mean level (M), and coupling tolerance (δ) by using a sufficient number of sections. The

4 TAPERED-LINE 90-DEGREE AND ASYMMETRIC COUPLERS

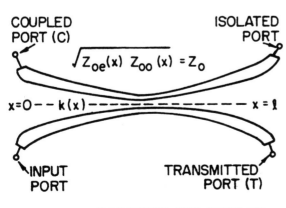

TAPERED SYMMETRIC 90° COUPLER

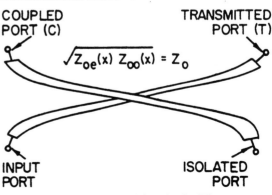

TAPERED 90° COUPLER WITH OVERLAPPED CONDUCTORS— STRIPLINE REALIZATION

Courtesy of R. Wenzel

A wide variety of coupler characteristics are available by using continuous coupling k(x) rather than stepped coupling. No exact design approach is available, but methods have been presented (8), (9) for achieving very precise designs. The use of a continuous coupling characteristic k(x) rather than a stepped characteristic appears to provide better directivity in very broadband couplers by reducing the discontinuity effects associated with an abrupt step in coupling. Both bandpass (8) and semi-infinite highpass (9) — (10) coupling characteristics can be obtained. For a given mean coupling level (M), coupling tolerance , and bandwidth B, tapered-line designs have been found to require a higher maximum coupling coefficient and slightly longer length than a comparable stepped coupling design. Multi-layer stripline construction with over-lapped center conductors has been utilized to achieve high values of k(x). A stripline over-lapped geometry can also be employed in stepped coupling designs. The full potential and limitations of tapered-line designs are not known at the present time, and further information on recent work can be found in (8) — (10).

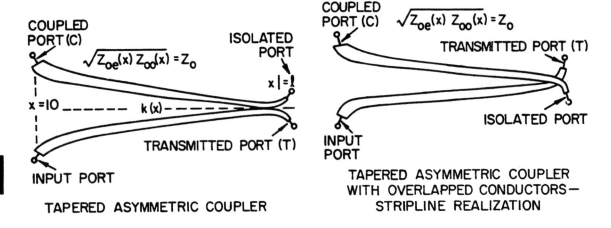

TAPERED ASYMMETRIC COUPLER

TAPERED ASYMMETRIC COUPLER
WITH OVERLAPPED CONDUCTORS—
STRIPLINE REALIZATION

5 180 DEGREE COUPLER, TAPERED LINE MAGIC T

A TEM coupler that performs as a Magic-T or 180-degree hybrid over a semi-infinite band (i.e., above a given cutoff frequency determined by the length of the coupler). The tapered coupled lines are designed to act as impedance transformers for the even- and odd-mode impedances (Z_{oe}, Z_{oo}) with coupler action due entirely to the reflections caused by the abrupt step between coupled and un-coupled lines. The 180-degree property of the device is not theoretically perfect, but is achieved above the cutoff frequency in an approximate manner and is dependent on the impedance taper utilized in the coupled-line region. For a given band of operation, the stripline Magic-T is often longer than 90-degree designs, and represents the only practical method demonstrated to date of achieving 180-degree hybrid performance over multi-octave frequency ranges in the microwave region. Similar performance can also be achieved using stepped coupled lines. For additional details, see (11).

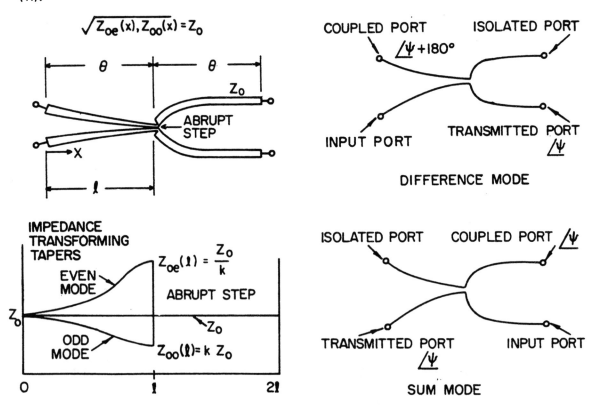

DIFFERENCE MODE

SUM MODE

Courtesy of R. Wenzel

6 TANDEM INTERCONNECTIONS

All of the multi-section and tapered-line couplers previously described can, in theory, be utilized to achieve any mean coupling level, coupling tolerance, and bandwidth. In practice, practical single structures can be realized for bandwidths of 10:1 and greater for low mean-coupling levels and typically for bandwidths of less than 10:1 for high mean-coupling values. A method of achieving bandwidths of 10:1 and greater in structures with practical coefficient of coupling values is to utilize a suitable tandem connection of couplers with lower mean-coupling level than that desired. For example, a 3-dB coupler can be obtained by a tandem interconnection of two 8.34-dB couplers. Many other tandem connections are possible and the component couplers do not have to be of equal complexity. The 90-degree or 180-degree property of the individual couplers is preserved in a proper tandem connection and either stepped coupling or tapered-line component devices can be utilized. An over-lapped stripline geometry is particularly desirable for tandem connections in that it allows a planar arrangement of ports to be easily achieved. Detailed discussions and design procedures for utilizing the tandem connection can be found in (11) - (13).

EXAMPLES OF TANDEM INTERCONNECTION

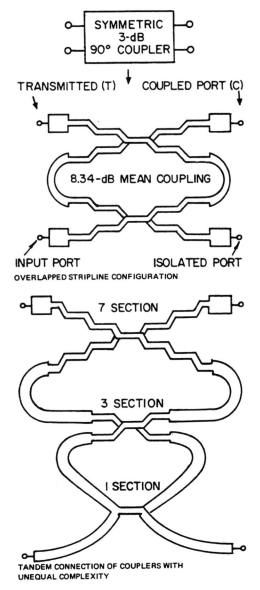

SYMMETRIC 3-dB 90° COUPLER

TRANSMITTED (T)   COUPLED PORT (C)

8.34-dB MEAN COUPLING

INPUT PORT   ISOLATED PORT

OVERLAPPED STRIPLINE CONFIGURATION

7 SECTION

3 SECTION

I SECTION

TANDEM CONNECTION OF COUPLERS WITH UNEQUAL COMPLEXITY

7 OTHER TEM COUPLERS – BRANCH LINE, HYBRID RING

Branch-line and hybrid-ring couplers are planar structures suitable for narrow or moderate bandwidth stripline or microstrip applications. Detailed design and performance data is well documented and can be found in several references (14) - (17). A new synthesis procedure and element value tables for branch-line couplers that gives exact results for maximally flat designs and near equal-ripple behavior for Chebyshev designs has recently been given (18).

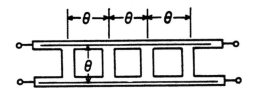

BRANCH LINE

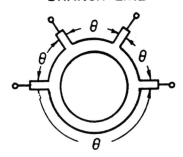

HYBRID RING

8 THREE-PORT HYBRIDS AS MATCHED IN-PHASE POWER DIVIDERS OR POWER COMBINERS

A three-port matched power divider with equal phase outputs or an equal phase power combiner. All ports are well matched with high isolation between output ports. Can be realized using stepped impedance lines and lumped resistors (19) or with tapered lines with a continuous resistive film (20). A practical structure for bandwidths of less than an octave to decade and greater. No exact synthesis procedure or tabular numerical data is available.

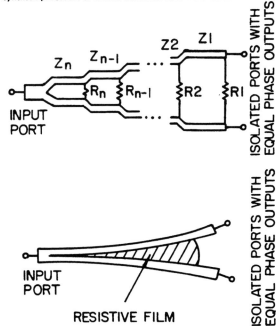

INPUT PORT

ISOLATED PORTS WITH EQUAL PHASE OUTPUTS

INPUT PORT

RESISTIVE FILM

ISOLATED PORTS WITH EQUAL PHASE OUTPUTS

REFERENCES

Note: Only a minimum of references is given. Many others are available with review articles and bibliographies to be found in (21) - (25).

1. E. G. Cristal and L. Young, "Tables of Optimum Symmetrical TEM-mode Coupled Transmission-Line Directional Couplers," IEEE TRANS. ON MICROWAVE THEORY AND TECHNIQUES, Vol. MTT-13, 1965, pp. 544-58.

2. P. P. Toulios and A. C. Todd, "Synthesis of Symmetrical TEM-mode Directional Couplers," IEEE TRANS. ON MICROWAVE THEORY AND TECHNIQUES, Vol. 13, 1965, pp. 536-44.

3. R. Levy, "General Synthesis of Asymmetric Multi-Element Coupled-Transmission-Line Directional Couplers," IEEE TRANS. ON MICROWAVE THEORY AND TECHNIQUES, Vol. MTT-11, 1963, pp. 226-37.

4. R. Levy, "Tables for Asymmetric Multi-Element Coupled Transmission Line Directional Couplers," IEEE TRANS. ON MICROWAVE THEORY AND TECHNIQUES, Vol. MTT-12, 1964, pp. 275-9.

5. D. I. Kraker, "Asymmetric Coupled-Transmission-Line Magic-T," IEEE TRANS. ON MICROWAVE THEORY AND TECHNIQUES, Vol. MTT-12, 1964, pp. 595-9.

6. S. B. Cohn, "The Re-Entrant Cross-Section and Wide-Band 3-dB Hybrid Couplers," IEEE TRANS. ON MICROWAVE THEORY AND TECHNIQUES, Vol. MTT-11, 1963, pp. 254-8.

7. R. Levy, "Transmission-Line Directional Couplers for Very Broad-Band Operation," PROCEEDINGS INST. ELEC. ENGRS. (London), Vol. 112, 1965, pp. 469-76.

8. C. P. Tresselt, "The Design and Construction of Broadband, High-Directivity, 90-Degree Couplers Using Nonuniform Line Techniques," IEEE TRANS. ON MICROWAVE THEORY AND TECHNIQUES, Vol. MTT-14, December, 1966, pp. 647-56.

9. F. Arndt, "High-Pass Transmission-Line Directional Coupler," IEEE TRANS. ON MICROWAVE THEORY AND TECHNIQUES, Vol. MTT-16, May, 1968, pp. 310-1.

10. C. B. Sharpe, "An Equivalence Principle For Nonuniform Transmission-Line Directional Couplers," IEEE TRANS. ON MICROWAVE THEORY AND TECHNIQUES, Vol. MTT-15, July, 1967, pp. 398-405.

11. R. H. DuHamel and M. E. Armstrong, "A Wide-Band Monopulse Antenna Utilizing the Tapered-Line Magic T," Fifteenth Annual Symposium, Air Force Avionics Laboratory, Wright-Patterson AFB, Ohio, in cooperation with the University of Illinois, Monticello, Illinois, October 12, 1965.

12. J. P. Shelton, Jr., J. J. Wolfe, and R. C. Van Wagoner, "Tandem Couplers and Phase Shifters for Multi-Octave Bandwidth," MICROWAVES, April, 1965, pp. 14-9.

13. J. P. Shelton and J. A. Mosko, "Synthesis and Design of Wide-Band Equal-Ripple TEM Directional Couplers and Fixed Phase Shifters," IEEE TRANS. ON MICROWAVE THEORY AND TECHNIQUES, Vol. MTT-14, October, 1966, pp. 264-73.

14. J. Reed and G. J. Wheeler, "A Method of Analysis of Symmetrical Four-Port Networks," IRE TRANS. ON MICROWAVE THEORY AND TECHNIQUES, Vol. 6, 1956, pp. 246-52.

15. G. L. Matthaei, L. Young, and E. M. T. Jones, Microwave Filters, Impedance Matching Networks, and Coupling Structures, (New York: McGraw-Hill), 1964.

16. W. A. Tyrell, "Hybrid Circuits for Microwaves," PROCEEDINGS OF THE IRE, Vol. 35, 1947, pp. 1307-13.

17. W. V. Tyminski and A. E. Hylas, "A Wide-Band Hybrid Ring for U.H.F.," PROCEEDINGS OF THE IRE, Vol. 41, 1953, pp. 81-7.

18. R. Levy and L. F. Lind, "Synthesis of Symmetrical Branch-Guide Directional Couplers," IEEE TRANS. ON MICROWAVE THEORY AND TECHNIQUES, Vol. MTT-16, February, 1968, pp. 80-9.

19. S. B. Cohn, "A Class of Broadband Three-Port TEM-Mode Hybrids," IEEE TRANS. ON MICROWAVE THEORY AND TECHNIQUES, Vol. MTT-16, February, 1968, pp. 110-6.

20. P. C. Goodman, "A Wideband Stripline Matched Power Divider," 1968 G-MTT International Microwave Symposium Digest, Detroit, Michigan, May 20-22, pp. 16-20.

21. Microwave Engineers' Handbook and Buyers' Guide, 1967, p. 117, and 1968, p. 210.

22. R. Levy, "Directional Couplers," Advances in Microwaves, (New York: Academic Press), 1966.

23. W. E. Caswell and R. F. Schwartz, "The Directional Coupler — 1966," IEEE TRANS. ON MICROWAVE THEORY AND TECHNIQUES (Correspondence), Vol. MTT-15, February, 1967, pp. 120-3.

24. L. Young, "Microwave Filters — 1965," IEEE TRANS. ON MICROWAVE THEORY AND TECHNIQUES, Vol. MTT-13, September, 1965, pp. 488-508.

25. L. Young, "Reflections on Microwave Filters and Couplers," Microwave Journal, April, 1968, pp. 54-8.

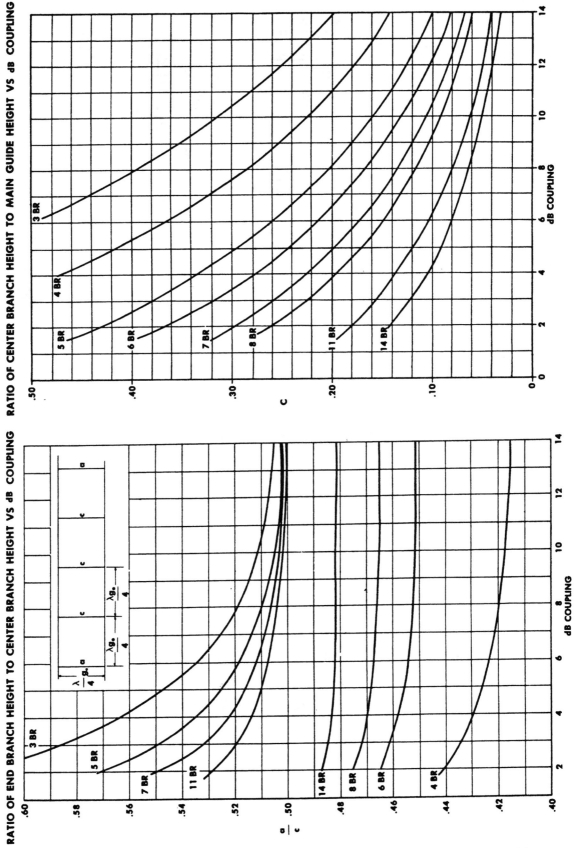

RATIO OF CENTER BRANCH HEIGHT TO MAIN GUIDE HEIGHT VS dB COUPLING

RATIO OF END BRANCH HEIGHT TO CENTER BRANCH HEIGHT VS dB COUPLING

John Reed, Raytheon Co., Wayland, Mass., from "Branch Waveguide Coupler Design Charts," the microwave journal, January 1963.

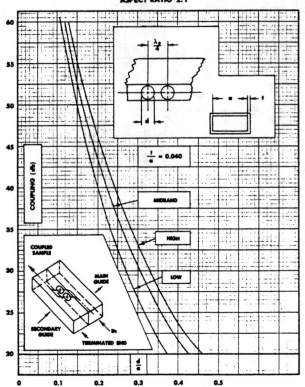

SIDEWALL DIRECTIONAL COUPLER
ASPECT RATIO 2:1

8

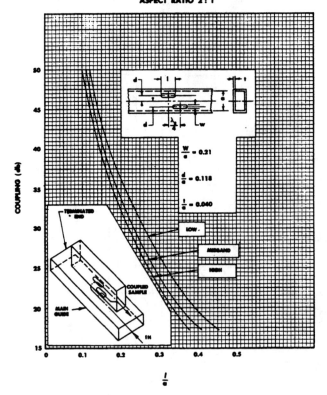

SCHWINGER REVERSED DIRECTIONAL COUPLER
ASPECT RATIO 2 : 1

Courtesy of Tore N. Anderson

MORENO CROSS GUIDE DIRECTIONAL COUPLER
ASPECT RATIO 2:1

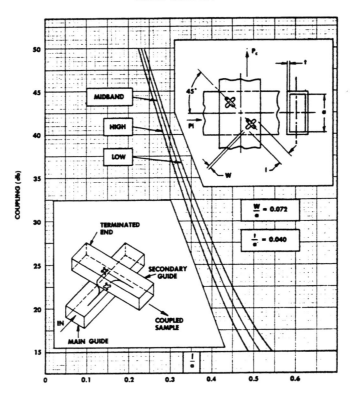

RIBLET TEE SLOT DIRECTIONAL COUPLER
ASPECT RATIO 2 : 1

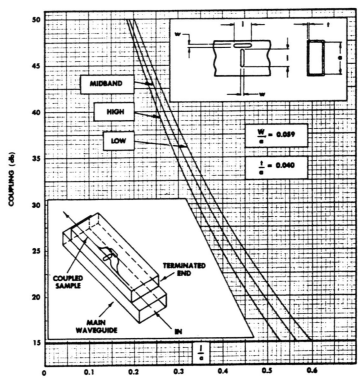

Courtesy of Tore N. Anderson.

BROADWALL DIRECTIONAL COUPLERS

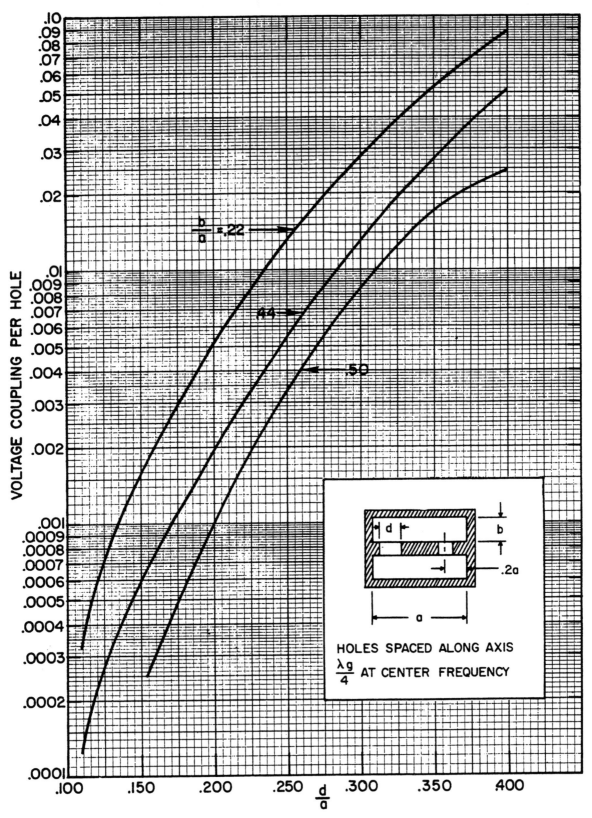

Courtesy of Gershon J. Wheeler

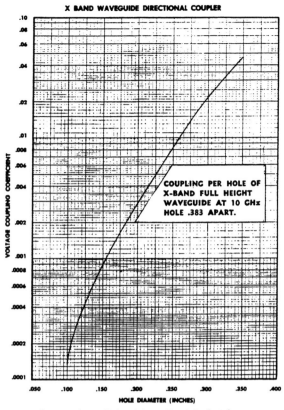

Courtesy of W. L. Shelton, Sylvania Electric Products, Inc.

ROUND HOLE CROSS GUIDE DIRECTIONAL COUPLER

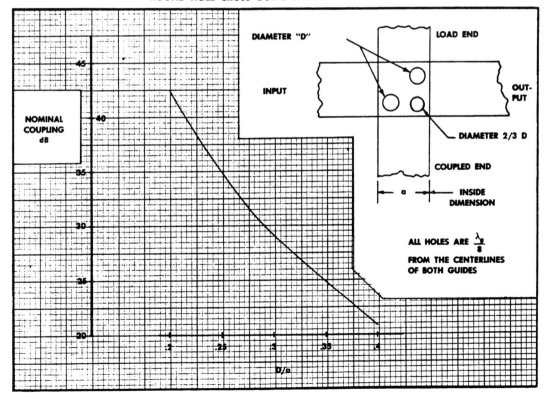

Courtesy of Gershon J. Wheeler.

12

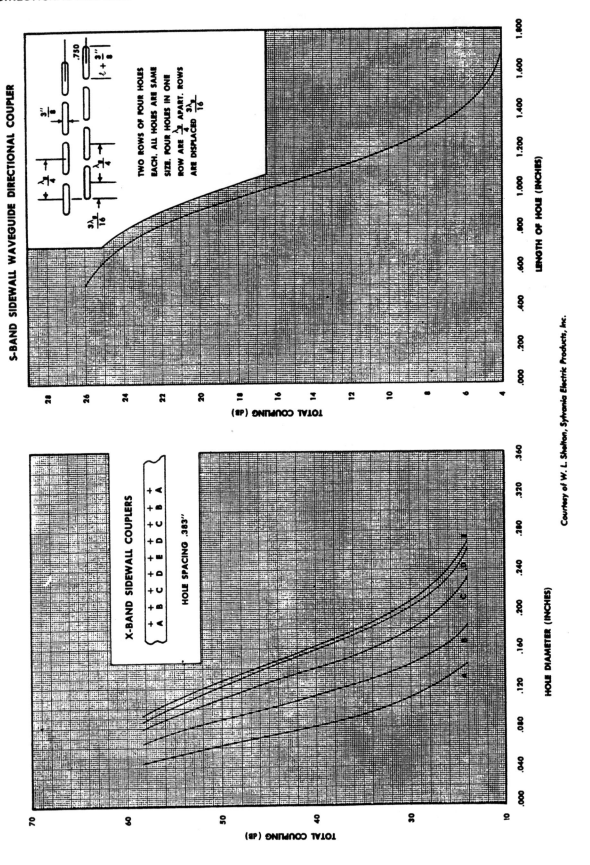

S-BAND SIDEWALL WAVEGUIDE DIRECTIONAL COUPLER

TWO ROWS OF FOUR HOLES EACH, ALL HOLES ARE SAME SIZE. FOUR HOLES IN ONE ROW ARE $\frac{\lambda}{4}$ APART. ROWS ARE DISPLACED $\frac{3\lambda}{16}$

LENGTH OF HOLE (INCHES)

TOTAL COUPLING (dB)

X-BAND SIDEWALL COUPLERS

HOLE SPACING .383"

A B C D E D C B A

HOLE DIAMETER (INCHES)

TOTAL COUPLING (dB)

Courtesy of W. L. Shelton, Sylvania Electric Products, Inc.

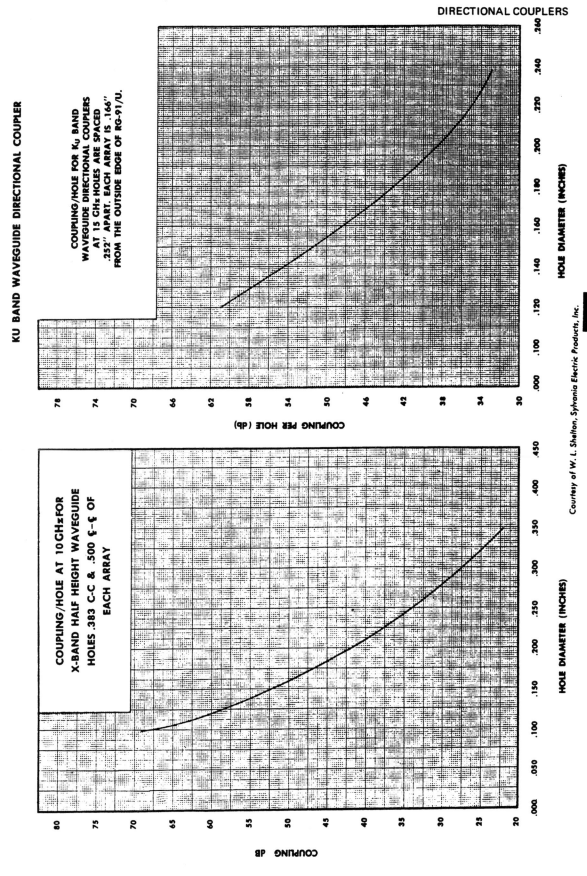

KU BAND WAVEGUIDE DIRECTIONAL COUPLER

COUPLING/HOLE FOR K$_U$ BAND WAVEGUIDE DIRECTIONAL COUPLERS AT 15 GHz HOLES ARE SPACED .252'' APART. EACH ARRAY IS .166'' FROM THE OUTSIDE EDGE OF RG-91/U.

COUPLING PER HOLE (db)

HOLE DIAMETER (INCHES)

COUPLING/HOLE AT 10GHzFOR X-BAND HALF HEIGHT WAVEGUIDE HOLES .383 C-C & .500 ¢–¢ OF EACH ARRAY

COUPLING dB

HOLE DIAMETER (INCHES)

Courtesy of W. L. Shelton, Sylvania Electric Products, Inc.

13

STRIP TRANSMISSION LINE DIRECTIONAL COUPLERS

w/b and w$_c$/b, ratios

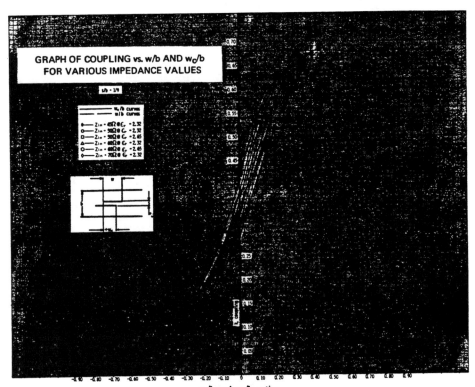

w/b and w$_c$/b, ratios

Courtesy of Joseph A. Mosko

14

STRIP TRANSMISSION LINE DIRECTIONAL COUPLERS

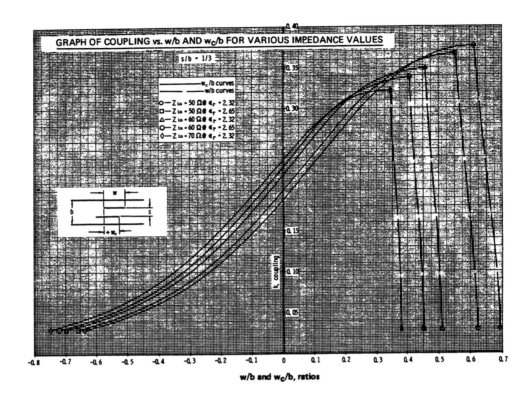

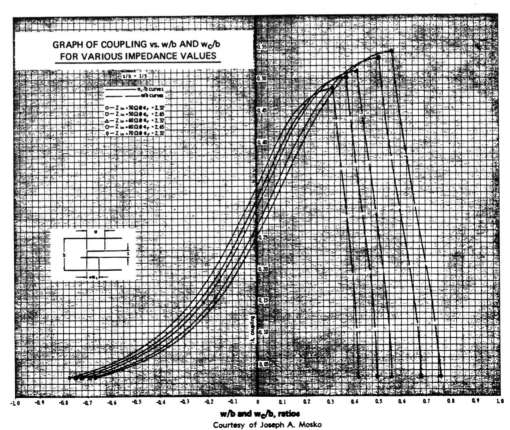

Courtesy of Joseph A. Mosko

ANTENNAS

CASSEGRAIN ANTENNAS—GEOMETRICAL RELATIONS

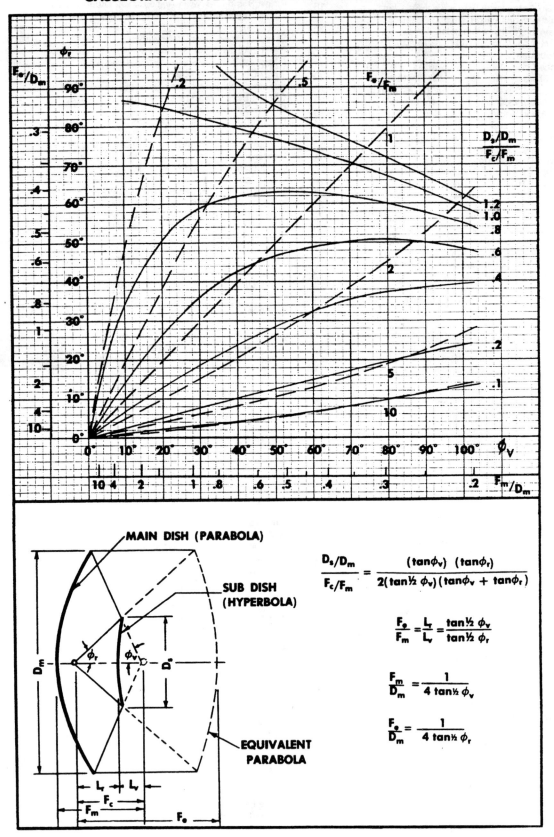

$$\frac{D_s/D_m}{F_c/F_m} = \frac{(\tan\phi_v)\,(\tan\phi_r)}{2(\tan\frac{1}{2}\,\phi_v)(\tan\phi_v + \tan\phi_r)}$$

$$\frac{F_e}{F_m} = \frac{L_r}{L_v} = \frac{\tan\frac{1}{2}\,\phi_v}{\tan\frac{1}{2}\,\phi_r}$$

$$\frac{F_m}{D_m} = \frac{1}{4\tan\frac{1}{2}\,\phi_v}$$

$$\frac{F_e}{D_m} = \frac{1}{4\tan\frac{1}{2}\,\phi_r}$$

Courtesy of Peter W. Hannan, Wheeler Laboratories, Smithtown, N. Y.

SPACE ATTENUATION VS. FEED ANGLE

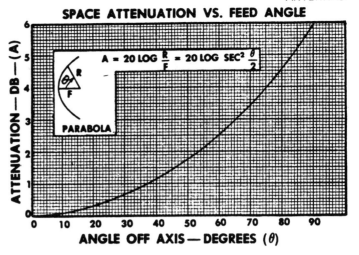

$$A = 20 \log \frac{R}{F} = 20 \log \sec^2 \frac{\theta}{2}$$

ATTENUATION—DB—(A)

ANGLE OFF AXIS—DEGREES (θ)

PARABOLIC ANTENNA

FEED DESIGN

UNIVERSAL FEED HORN PATTERN

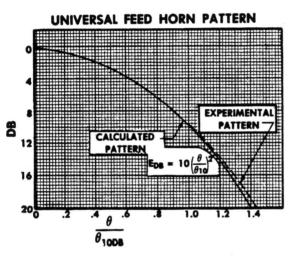

EXPERIMENTAL PATTERN

CALCULATED PATTERN

$$E_{DB} = 10 \left(\frac{\theta}{\theta_{10}} \right)^2$$

DB

$\dfrac{\theta}{\theta_{10DB}}$

TO USE:
1. Start with aperture edge illumination & f/D.
2. Find feed angle using f/D.
3. Find space attenuation using feed angle.
4. Subtract space attenuation from desired edge illumination to get feed taper.
5. Find feed 10 db beamwidth using feed taper.

19

f/D VS. SUBTENDED ANGLE AT FOCUS

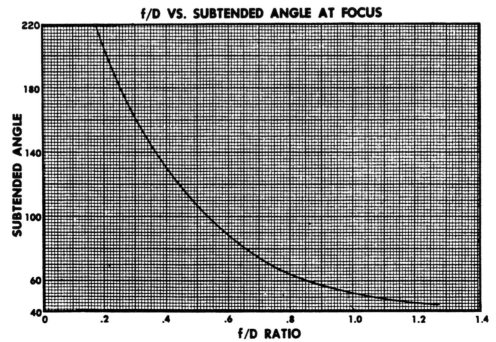

SUBTENDED ANGLE

f/D RATIO

Courtesy of K. S. Kelleher, Aero Geo Astro Corp., Alexandria, Va.

Dish Gain vs Aperture Diameter

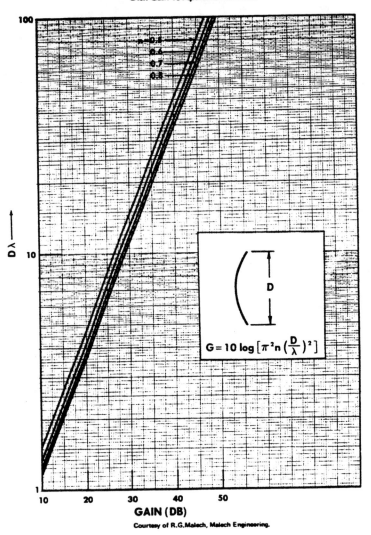

$$G = 10 \log \left[\pi^2 n \left(\frac{D}{\lambda} \right)^2 \right]$$

GAIN (DB)

Courtesy of R.G. Malech, Malech Engineering.

Gain Factor for Circular Aperture Taylor Distributions

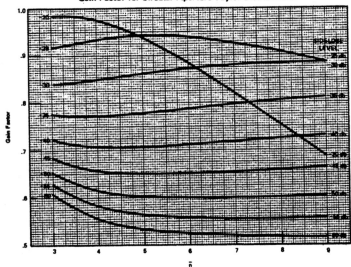

Courtesy of A.W. Love, Autonetics Division, North American Rockwell, Anaheim, California.

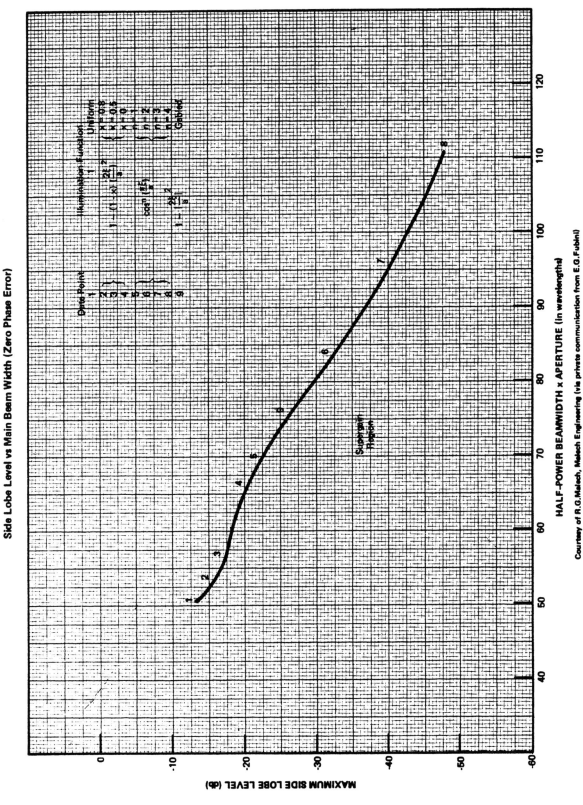

Side Lobe Level vs Main Beam Width (Zero Phase Error)

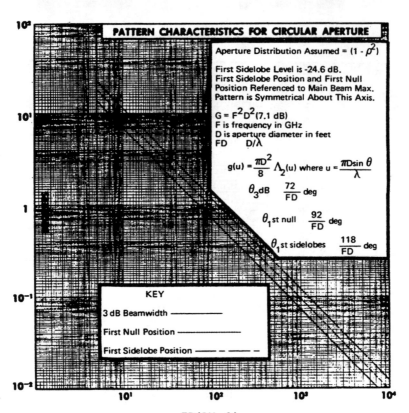

PATTERN CHARACTERISTICS FOR CIRCULAR APERTURE

Aperture Distribution Assumed = $(1 - \rho^2)$

First Sidelobe Level is -24.6 dB.
First Sidelobe Position and First Null
Position Referenced to Main Beam Max.
Pattern is Symmetrical About This Axis.

$G = F^2 D^2 (7.1 \text{ dB})$
F is frequency in GHz
D is aperture diameter in feet
FD D/λ

$g(u) = \frac{\pi D^2}{8} \Lambda_2(u)$ where $u = \frac{\pi D \sin \theta}{\lambda}$

θ_{3dB} $\frac{72}{FD}$ deg

θ_1st null $\frac{92}{FD}$ deg

θ_1st sidelobes $\frac{118}{FD}$ deg

KEY

3 dB Beamwidth ——————

First Null Position ————————

First Sidelobe Position — — — — —

FD (GHz, ft)

Courtesy of Stanley R. Jones, Staff Consultant, Aero Geo Astro Corp.

PARABOLOIDAL DIRECTIVITY LOSS DUE TO RANDOM ERRORS

STANDARD DEVIATION OF REFLECTOR ERROR, IN RADIANS

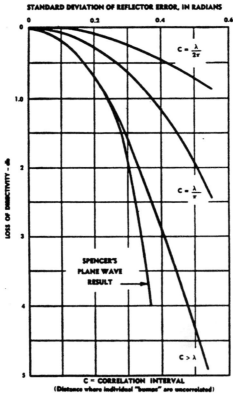

$C = \frac{\lambda}{2\pi}$

$C = \frac{\lambda}{\pi}$

SPENCER'S
PLANE WAVE
RESULT →

$C > \lambda$

LOSS OF DIRECTIVITY - dB

C = CORRELATION INTERVAL
(Distance where individual "bumps" are uncorrelated)

Reprinted from Microwave Scanning Antennas, edited by R. C. Hansen, Vol. I, p. 78. Copyright 1964 by Academic Press Inc., New York. Courtesy of Dr. J. Ruze.

Aperture Gain Loss Due to Random Errors

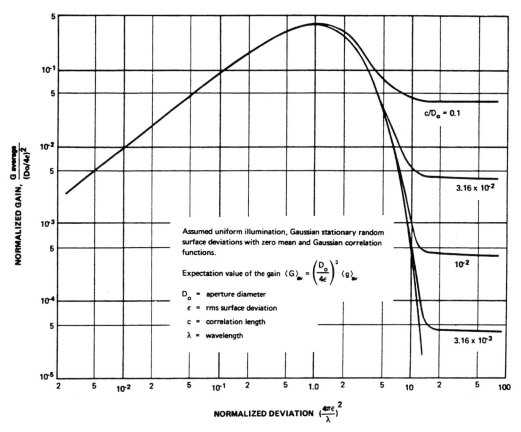

Reprinted from the Bell System Technical Journal, Vol. 47, No. 8, October, 1968, pg. 1643.

23

E and H Plane Sidelobes vs H Plane Taper

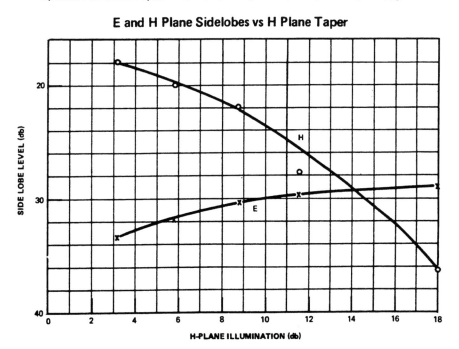

Reprinted from Antenna Engineering Handbook, edited by Henry Jasik, copyright 1961 by McGraw - Hill Book Co., Inc.

Beamwidth Conversion

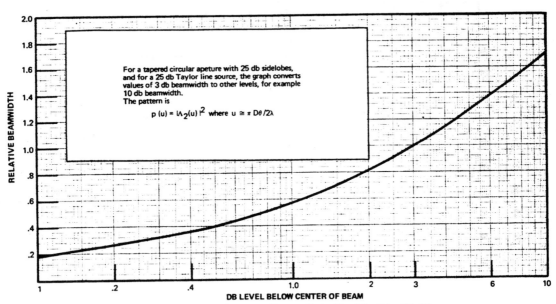

For a tapered circular aperture with 25 db sidelobes, and for a 25 db Taylor line source, the graph converts values of 3 db beamwidth to other levels, for example 10 db beamwidth.
The pattern is

$$p(u) = |\Lambda_2(u)|^2 \quad \text{where } u \cong \pi D\theta/2\lambda$$

RELATIVE BEAMWIDTH

DB LEVEL BELOW CENTER OF BEAM

Courtesy of George Stern, James Eickmann, Hughes Aircraft Co., Culver City, Cal.

24

Sidelobe & Gain Degradation of Blocked Aperture

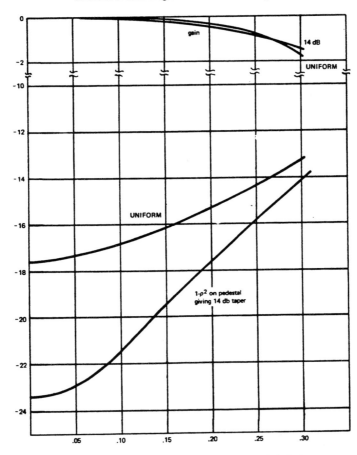

gain

14 dB

UNIFORM

UNIFORM

$1-\rho^2$ on pedestal giving 14 db taper

Courtesy of Dr. J.W.M.Baars, Radioobservatory, Dwingeloo, Netherlands.

Beam Deviation After Ruze

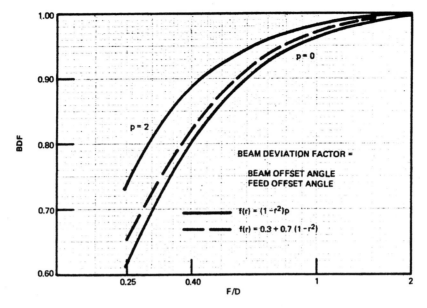

BEAM DEVIATION FACTOR.

Courtesy of Dr. John Ruze, Lincoln Labs., Reference: Transactions of IEEE, Vol. AP - B,
September 1965, pp. 660 - 665.

25

Directivity Degradation After Silver

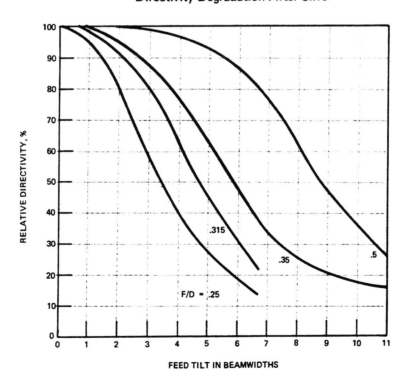

Reprinted from Microwave Antenna Theory and Design, MIT Rad. Lab. Series,
Vol. 12, edited by Samuel Silver, Copyright 1949 by McGraw - Hill Book Co., Inc.

Optimum Angular Accuracy of Circular Lobe Comparison Antenna

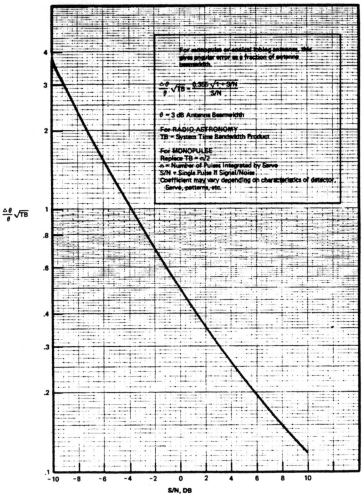

Courtesy of R. Manasse, Maximum Angular Accuracy of Tracking a Radio Star by Lobe Comparison, IRE Trans. on Antennas and Propagation, Vol. AP - B, No. 1, Jan., 1960.

Flyswatter Reflector Directivity

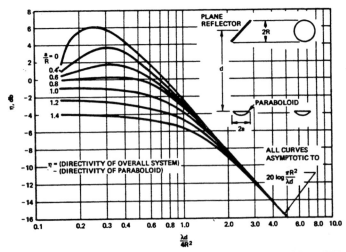

Reprinted from Antenna Engineering Handbook, edited by Henry Jasik, copyright 1961 by McGraw - Hill Book Co., Inc.

Offset Paraboloid Feed Design

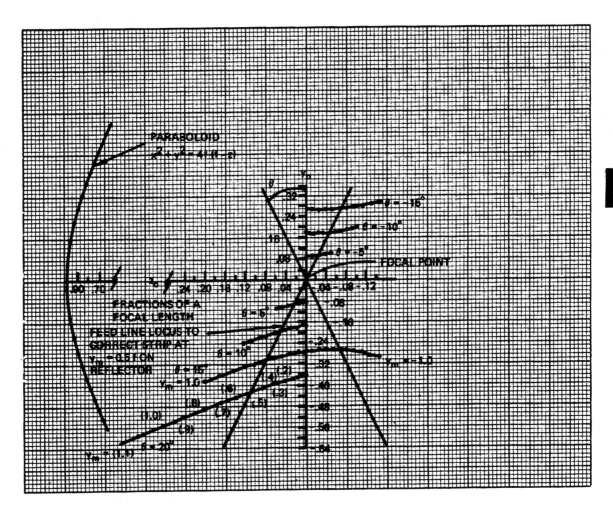

Prepared for Antenna Theory Part II by C. J. Sletten, Air Force Cambridge Laboratories, L. G. Hanscom Field, Bedford, Mass. To be published by McGraw-Hill Book Co.

Offset Paraboloid Feed Loci.

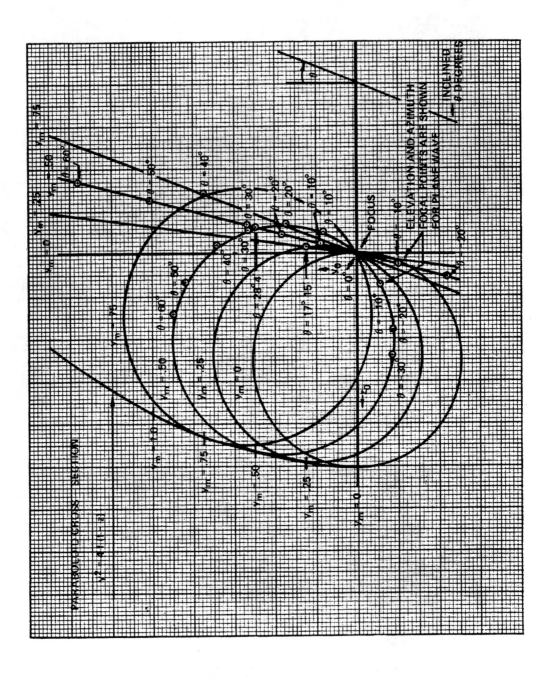

Prepared for Antenna Theory Part II by C. J. Sletten, Air Force Cambridge Laboratories,
L. G. Hanscom Field, Bedford, Mass. To be published by McGraw-Hill Book Co.

Feed Length For Spherical Reflector With Correcting Line Source

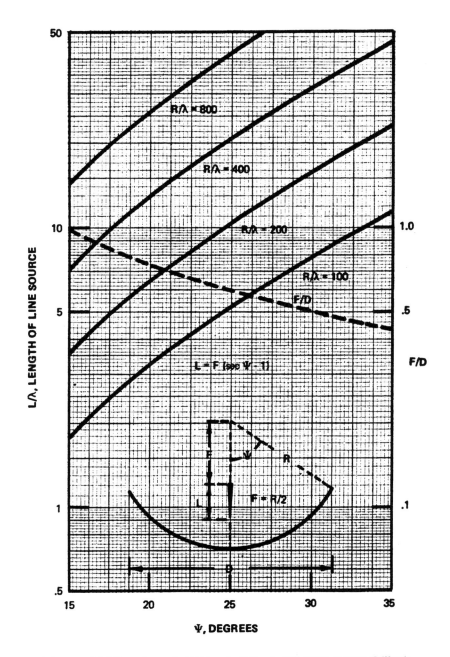

Courtesy of A.W. Love, Autonetics Division, North American Rockwell, Anaheim, California.

Bandwidth for Spherical Reflector with Correcting Line Source Feed

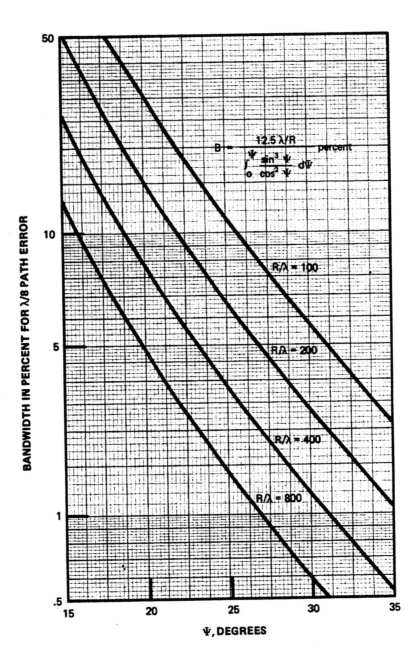

$$B = \frac{12.5\,\lambda/R}{\int_0^\Psi \frac{\sin^3\Psi}{\cos^2\Psi}\,d\Psi} \quad \text{percent}$$

BANDWIDTH IN PERCENT FOR λ/8 PATH ERROR

Ψ, DEGREES

R/λ = 100

R/λ = 200

R/λ = 400

R/λ = 800

Courtesy of A.W.Love, Autonetics Division, North American Rockwell, Anaheim, California.

Wind Forces on Standard Antennas with Radomes

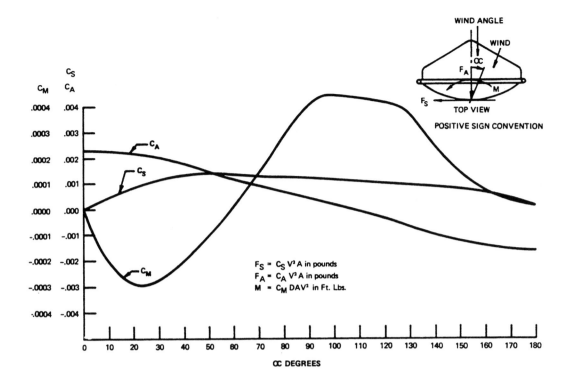

$F_S = C_S V^2 A$ in pounds
$F_A = C_A V^2 A$ in pounds
$M = C_M DAV^2$ in Ft. Lbs.

POSITIVE SIGN CONVENTION

Wind Forces on Standard Antennas

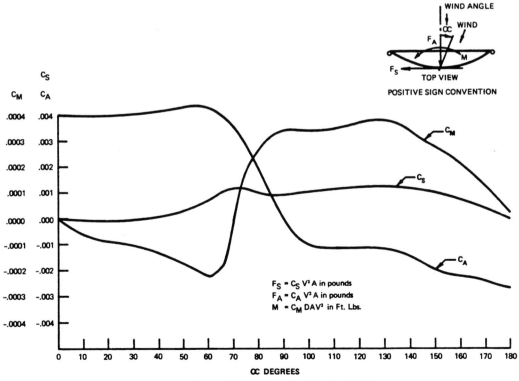

$F_S = C_S V^2 A$ in pounds
$F_A = C_A V^2 A$ in pounds
$M = C_M DAV^2$ in Ft. Lbs.

POSITIVE SIGN CONVENTION

Courtesy of Andrew Corporation, Orland Park, Illinois.

FAR FIELD DISTANCE FOR TYPICAL PARABOLIC ANTENNA
FOR USE IN COMPUTING POWER DENSITY VALUES

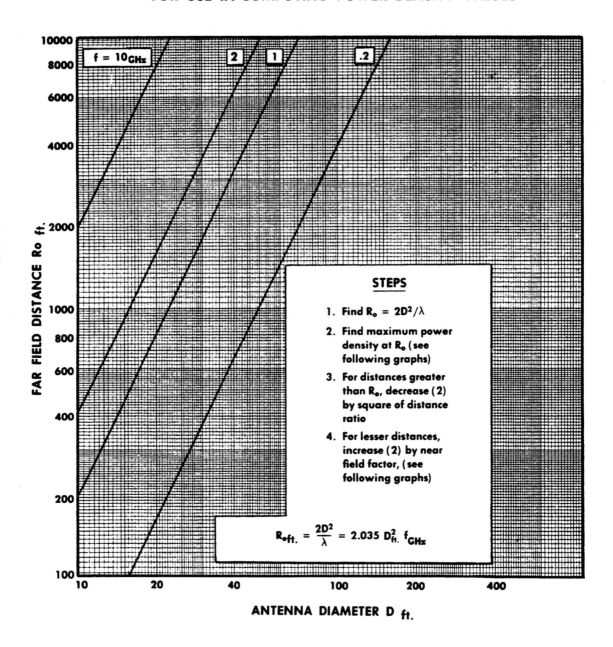

STEPS

1. Find $R_o = 2D^2/\lambda$

2. Find maximum power density at R_o (see following graphs)

3. For distances greater than R_o, decrease (2) by square of distance ratio

4. For lesser distances, increase (2) by near field factor, (see following graphs)

$$R_{o\,ft.} = \frac{2D^2}{\lambda} = 2.035 \, D_{ft.}^2 \, f_{GHz}$$

Courtesy of R. C. Hansen

Power Density at 2D²/λ For Typical Antenna

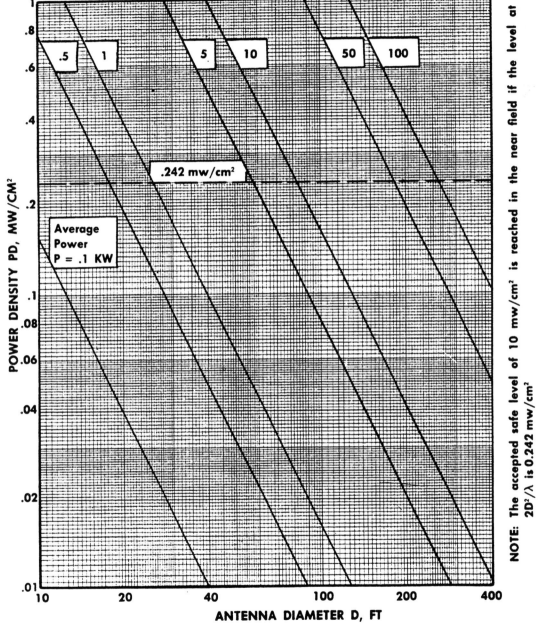

NOTE: The accepted safe level of 10 mw/cm² is reached in the near field if the level at 2D²/λ is 0.242 mw/cm²

$$PD = \frac{3\pi P}{64D^2} = \frac{158.4\,P_{KW}}{D^2\,ft} \quad mw/cm^2 \text{ (tapered illumination)}$$

33

Courtesy of R. C. Hansen

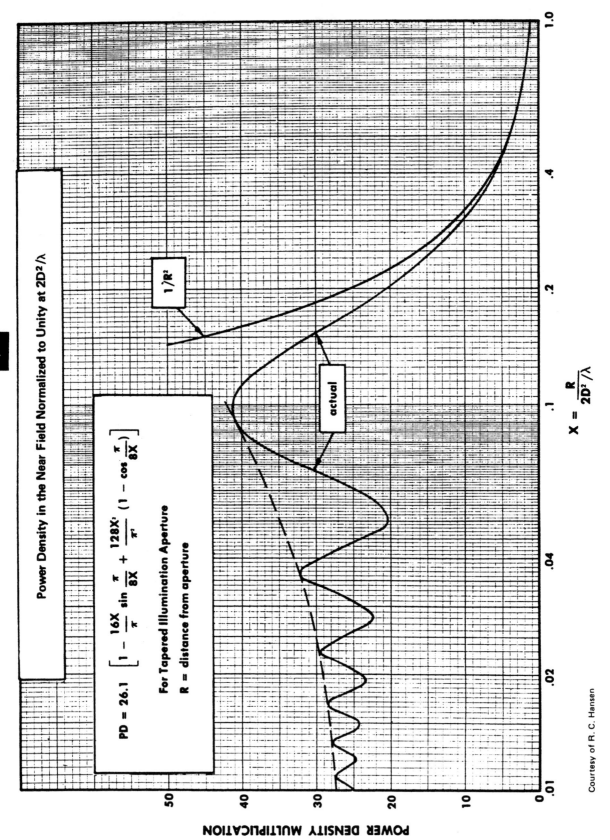

Power Density in the Near Field Normalized to Unity at $2D^2/\lambda$

$$PD = 26.1 \left[1 - \frac{16X}{\pi} \sin \frac{\pi}{8X} + \frac{128X^2}{\pi^2} \left(1 - \cos \frac{\pi}{8X} \right) \right]$$

For Tapered Illumination Aperture

R = distance from aperture

$1/R^2$

actual

$X = \dfrac{R}{2D^2/\lambda}$

POWER DENSITY MULTIPLICATION

Courtesy of R. C. Hansen

Uniform Line Source Power Density in the Near Field

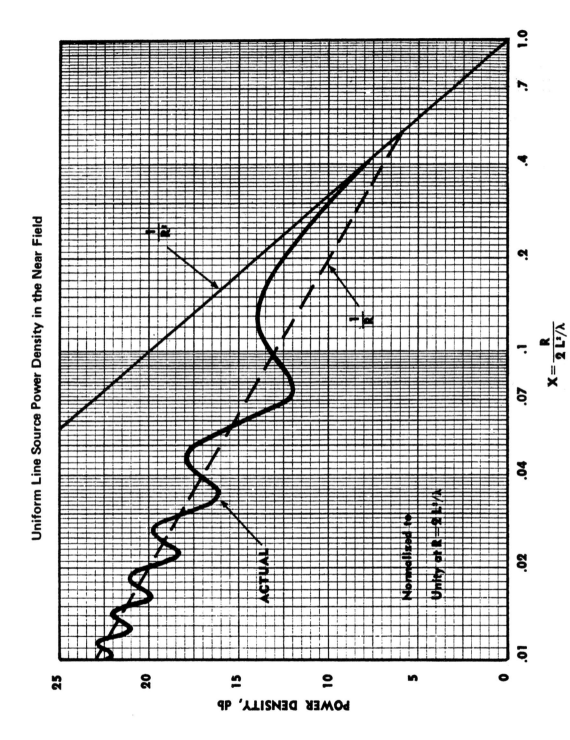

Reprinted from Microwave Scanning Antennas, edited by R.C.Hansen, Vol. 1, p. 36, copyright 1984 by Academic Press, Inc., New York. Courtesy of L.J.Ricardi.

35

36

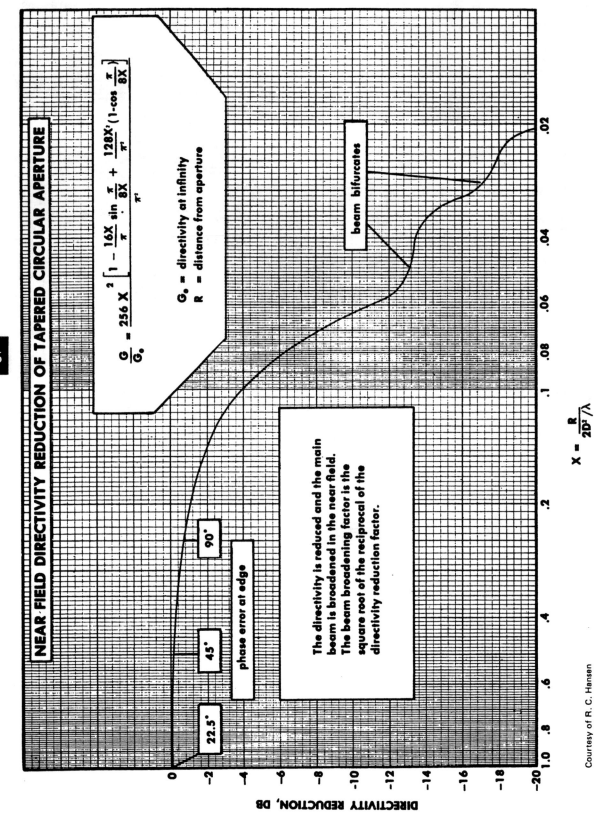

NEAR-FIELD DIRECTIVITY REDUCTION OF TAPERED CIRCULAR APERTURE

$$\frac{G}{G_\infty} = 256 \, X^2 \left[1 - \frac{16X}{\pi} \sin \frac{\pi}{8X} + \frac{128X^2}{\pi^2} \left(1 - \cos \frac{\pi}{8X} \right) \right]$$

G_∞ = directivity at infinity

R = distance from aperture

beam bifurcates

90°

45°

22.5°

phase error at edge

The directivity is reduced and the main beam is broadened in the near field. The beam broadening factor is the square root of the reciprocal of the directivity reduction factor.

$$X = \frac{R}{2D^2/\lambda}$$

DIRECTIVITY REDUCTION, DB

Courtesy of R. C. Hansen

Near Field Directivity Reduction of Rectangular Aperture With Separable Illumination

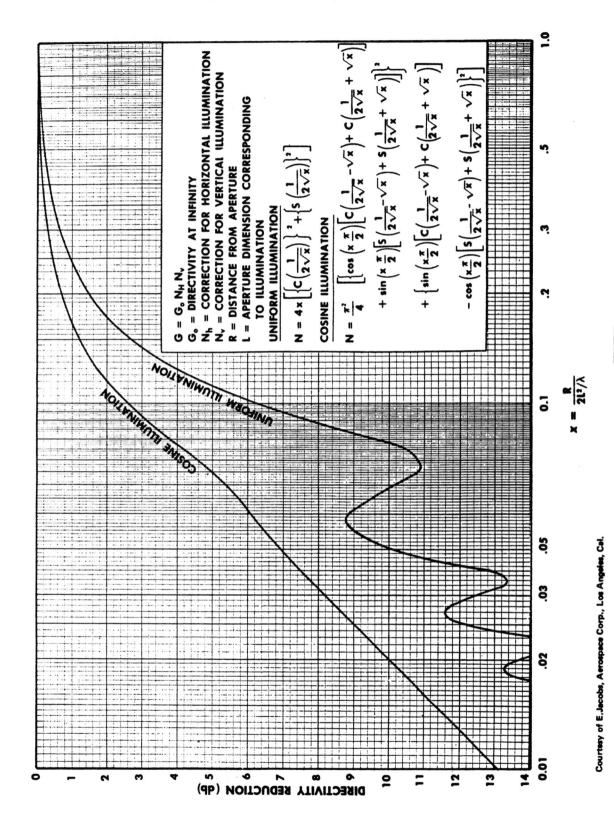

$G = G_o N_h N_v$

G_o = DIRECTIVITY AT INFINITY

N_h = CORRECTION FOR HORIZONTAL ILLUMINATION

N_v = CORRECTION FOR VERTICAL ILLUMINATION

R = DISTANCE FROM APERTURE

L = APERTURE DIMENSION CORRESPONDING TO ILLUMINATION

UNIFORM ILLUMINATION

$$N = 4x \left[\left\{ C\left(\frac{1}{2\sqrt{x}}\right) \right\}^2 + \left\{ S\left(\frac{1}{2\sqrt{x}}\right) \right\}^2 \right]$$

COSINE ILLUMINATION

$$N = \frac{\pi^2}{4} \Bigg\{ \left[\cos\left(x\frac{\pi}{2}\right) \left[C\left(\frac{1}{2\sqrt{x}} - \sqrt{x}\right) + C\left(\frac{1}{2\sqrt{x}} + \sqrt{x}\right) \right] \right. $$
$$+ \sin\left(x\frac{\pi}{2}\right) \left[S\left(\frac{1}{2\sqrt{x}} - \sqrt{x}\right) + S\left(\frac{1}{2\sqrt{x}} + \sqrt{x}\right) \right] \Bigg]^2 $$
$$+ \Bigg\{ \sin\left(x\frac{\pi}{2}\right) \left[C\left(\frac{1}{2\sqrt{x}} - \sqrt{x}\right) + C\left(\frac{1}{2\sqrt{x}} + \sqrt{x}\right) \right] $$
$$- \cos\left(x\frac{\pi}{2}\right) \left[S\left(\frac{1}{2\sqrt{x}} - \sqrt{x}\right) + S\left(\frac{1}{2\sqrt{x}} + \sqrt{x}\right) \right] \Bigg\}^2 \Bigg\}$$

UNIFORM ILLUMINATION

COSINE ILLUMINATION

$x = \dfrac{R}{2L^2/\lambda}$

DIRECTIVITY REDUCTION (dB)

Courtesy of E. Jacobs, Aerospace Corp., Los Angeles, Cal.

37

38

Minimum Spot Size of Focused Aperture

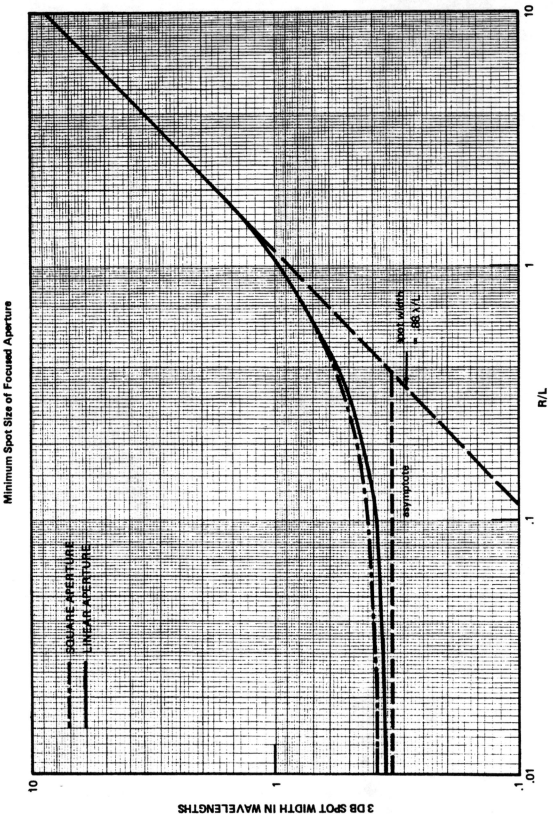

Reprinted from Minimum Spot Size of Focused Aperture, R.C.Hansen, Proc. U.S.R.I. Symp. on Electromagnetic Wace Theory, Delft, Netherlands, September 1965.

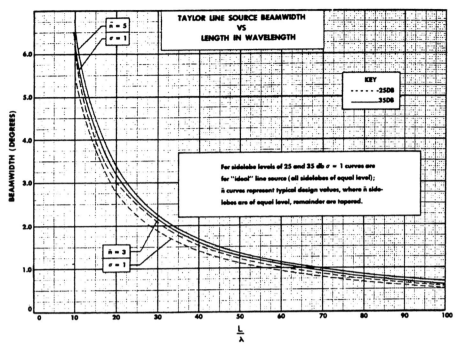

Courtesy of George Stern, James Eickmann, Hughes Aircraft Co., Culver City, Cal.

HIGHEST SIDELOBE OF LINE SOURCE WITH CENTER GAP

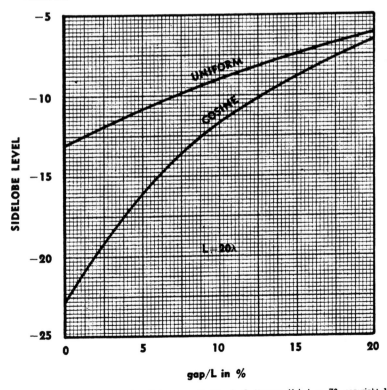

Reprinted from *Microwave Scanning Antennas*, edited by R. C. Hansen, Vol. I, p. 72, copyright 1964 by Academic Press Inc., New York. Courtesy of R. A. Gerlock.

40

APPROXIMATE ELECTRIC FIELD FOR BURIED OR SUBMERGED ANTENNA

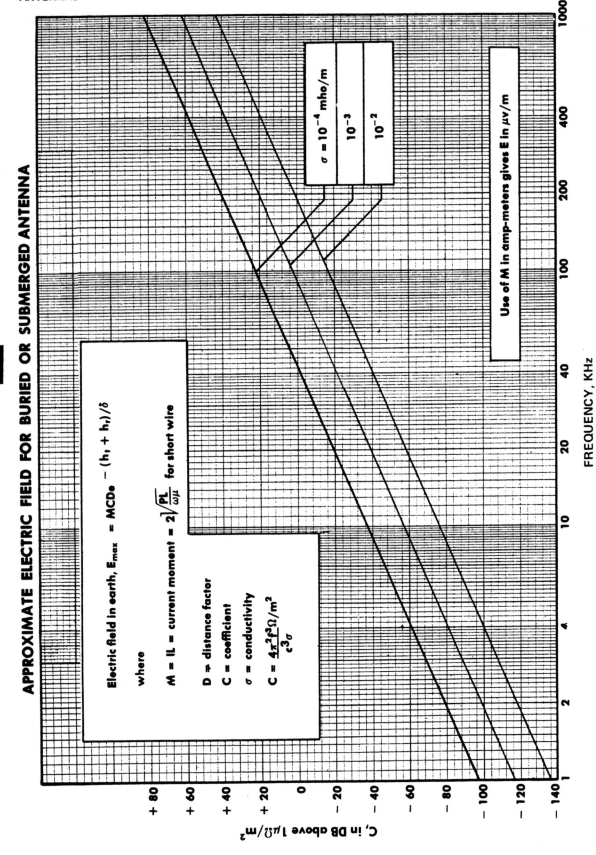

Electric field in earth, $E_{max} = MCDe^{-(h_t + h_r)/\delta}$

where

$M = IL =$ current moment $= 2\sqrt{\dfrac{PL}{\omega\mu}}$ for short wire

$D =$ distance factor

$C =$ coefficient

$\sigma =$ conductivity

$C = \dfrac{4\pi^2 f^3 \Omega/m^2}{c^3 \sigma}$

$\sigma = 10^{-4}$ mho/m

10^{-3}

10^{-2}

Use of M in amp-meters gives E in $\mu v/m$

FREQUENCY, KHz

C, in DB above $1\mu\Omega/m^2$

Courtesy of R. C. Hansen

NORMALIZED DISTANCE FUNCTION FOR BURIED ANTENNA

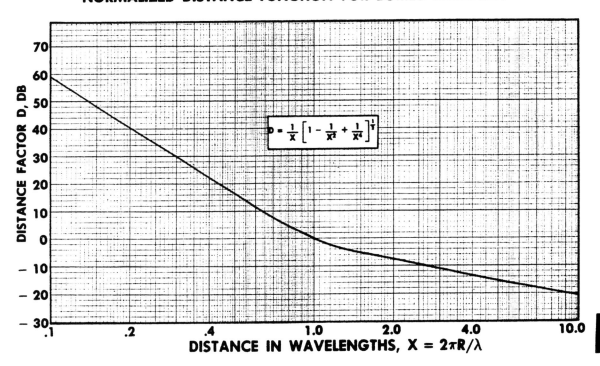

$$D = \frac{1}{X}\left[1 - \frac{1}{X^2} + \frac{1}{X^4}\right]^{\frac{1}{2}}$$

DISTANCE IN WAVELENGTHS, $X = 2\pi R/\lambda$

DEPTH ATTENUATION FOR BURIED ANTENNA

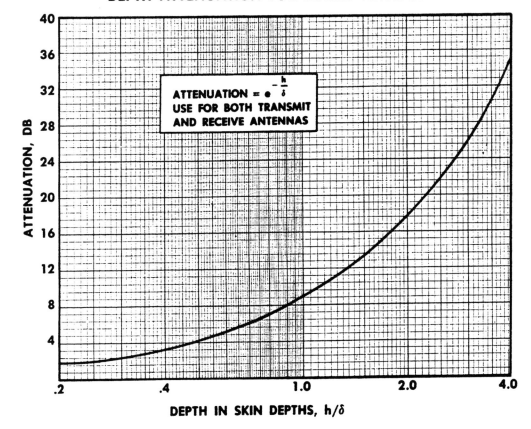

$$\text{ATTENUATION} = e^{-\frac{h}{\delta}}$$
USE FOR BOTH TRANSMIT AND RECEIVE ANTENNAS

DEPTH IN SKIN DEPTHS, h/δ

Courtesy of R. C. Hansen

Handbook - Volume Two

SKIN DEPTH IN EARTH OR SEA AT VLF & LF

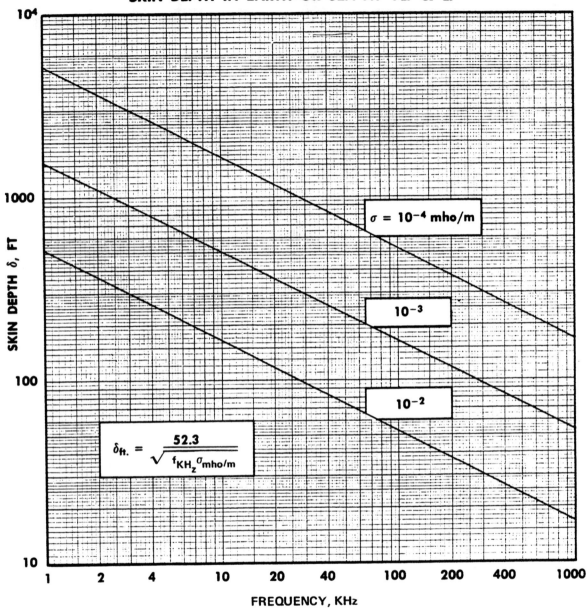

Courtesy of R. C. Hansen

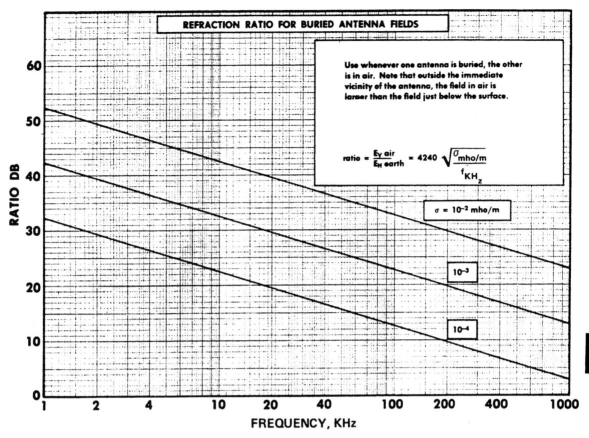

REFRACTION RATIO FOR BURIED ANTENNA FIELDS

Use whenever one antenna is buried, the other is in air. Note that outside the immediate vicinity of the antenna, the field in air is larger than the field just below the surface.

$$\text{ratio} = \frac{E_V \text{ air}}{E_H \text{ earth}} = 4240 \sqrt{\frac{\sigma_{\text{mho/m}}}{f_{\text{KH}_z}}}$$

$\sigma = 10^{-2}$ mho/m

10^{-3}

10^{-4}

RATIO DB

FREQUENCY, KHz

43

CONVERSION OF FIELD STRENGTH IN μ V/M
TO db / K T₀ B FOR SHORT DIPOLE

USE FOR CONVERTING NOISE DATA

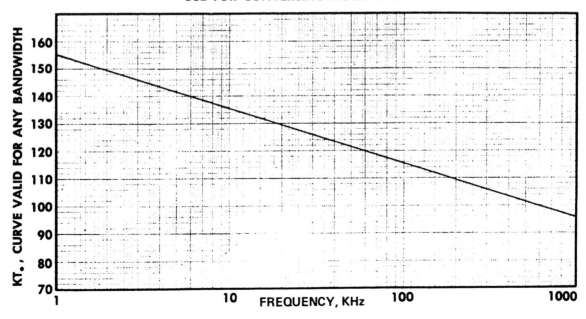

KT₀, CURVE VALID FOR ANY BANDWIDTH

FREQUENCY, KHz

44

RELATIVE VOLTAGE GAIN OF LOOP
WITH HOLLOW PROLATE SPHEROIDAL CORE

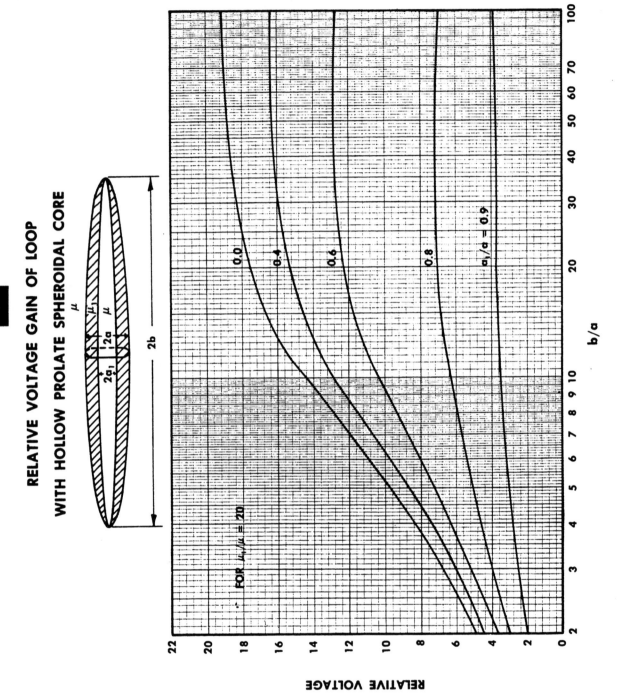

James R. Wait, Reproduced by courtesy of Canadian Journal of Technology.

RELATIVE VOLTAGE GAIN OF LOOP WITH HOLLOW PROLATE SPHEROIDAL CORE

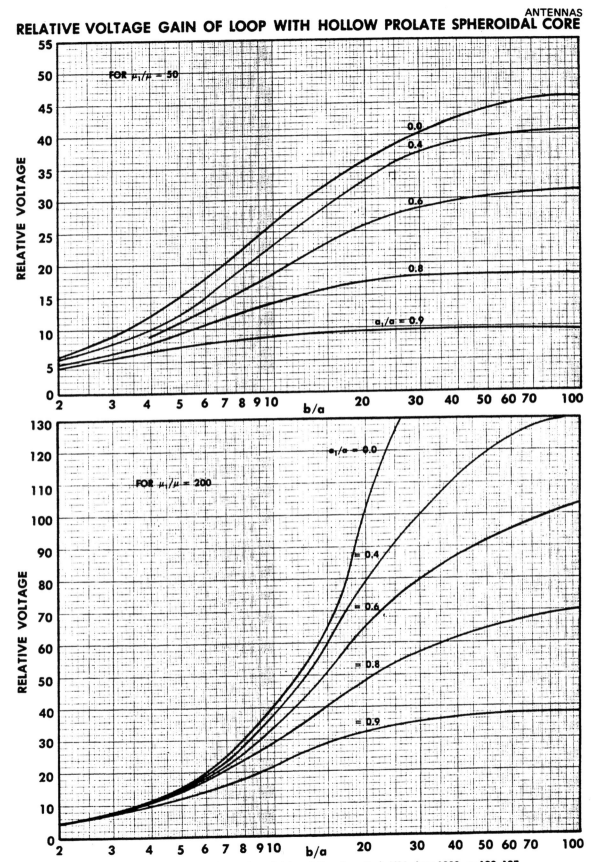

Ref.: "The Receiving Loop with a Hollow Prolate Spheroidal Core," Can. Jour. Tech. V31, June 1953, pp 132–137.
James R. Wait, Reproduced by courtesy of Canadian Journal of Technology.

46

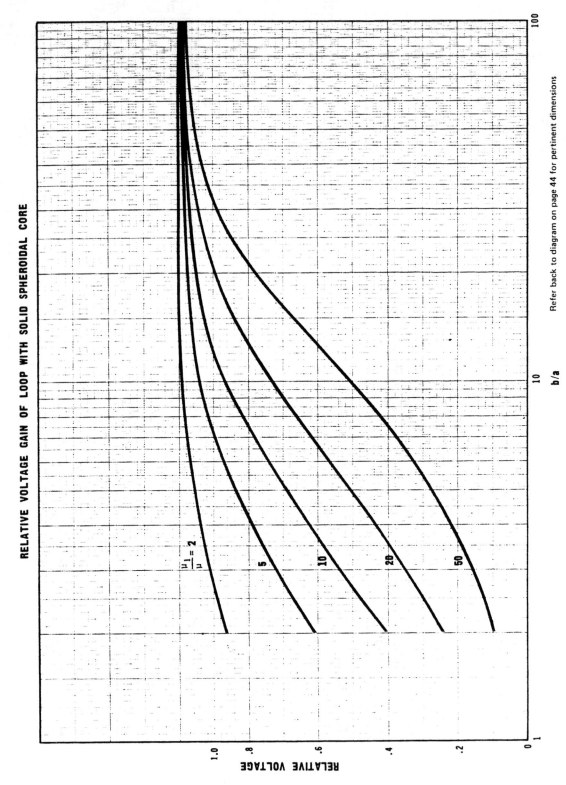

RELATIVE VOLTAGE GAIN OF LOOP WITH SOLID SPHEROIDAL CORE

James R. Wait, Reproduced by courtesy of Canadian Journal of Technology

Refer back to diagram on page 44 for pertinent dimensions

VELOCITY RATIO FOR MAXIMUM GAIN SURFACE WAVE ANTENNAS

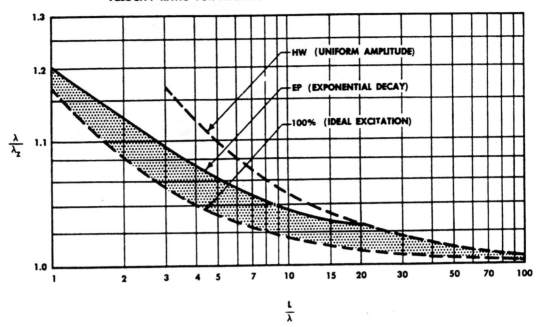

Reprinted from Antenna Engineering Handbook, edited by Henry Jasik, copyright 1961 by McGraw-Hill Book Co., Inc.

DIRECTIVITY AND BEAMWIDTH OF SURFACE WAVE ANTENNAS

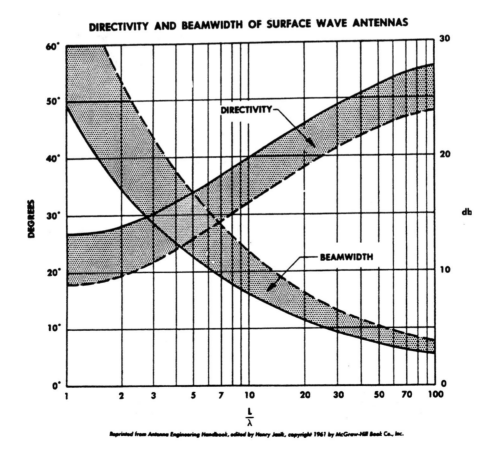

Reprinted from Antenna Engineering Handbook, edited by Henry Jasik, copyright 1961 by McGraw-Hill Book Co., Inc.

OPTIMUM RECTANGULAR HORNS, GIVING FLARE ANGLE vs LENGTH, WITH BEAMWIDTH ON AN AUXILIARY SCALE

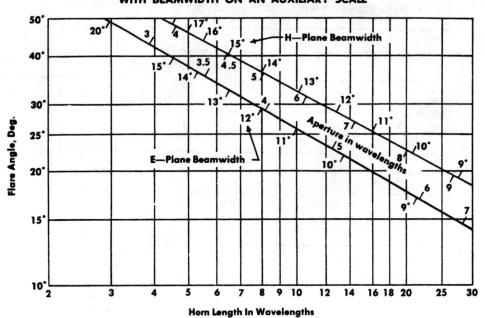

48

AXIAL MODE HELICAL ANTENNA BEAMWIDTH vs LENGTH

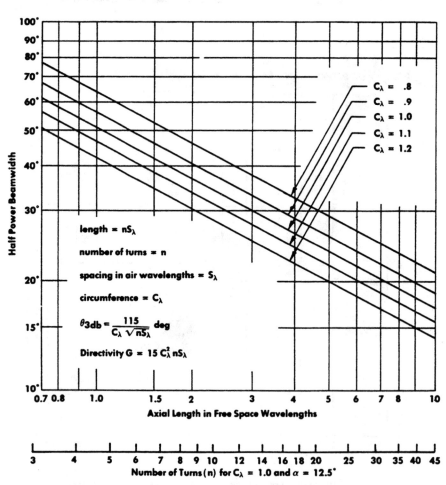

length = nS_λ

number of turns = n

spacing in air wavelengths = S_λ

circumference = C_λ

$$\theta_{3db} = \frac{115}{C_\lambda \sqrt{nS_\lambda}} \text{ deg}$$

Directivity G = $15 C_\lambda^2 nS_\lambda$

Number of Turns (n) for $C_\lambda = 1.0$ and $\alpha = 12.5°$

Courtesy of J. D. Kraus, Antennas, copyright 1950, McGraw-Hill Book Co.

49

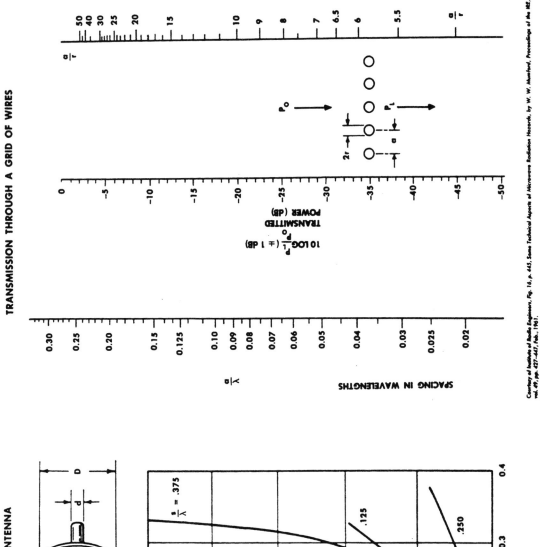

TRANSMISSION THROUGH A GRID OF WIRES

Courtesy of Institute of Radio Engineers, Fig. 16, p. 445, Some Technical Aspects of Microwave Radiation Hazards, by W. W. Mumford, Proceedings of the IRE, vol. 49, pp. 427-447, Feb. 1961.

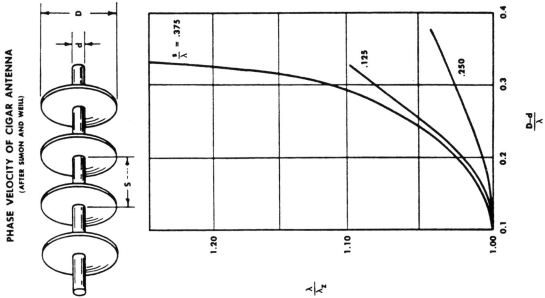

PHASE VELOCITY OF CIGAR ANTENNA
(AFTER SIMON AND WEILL)

Reprinted from Antenna Engineering Handbook, edited by Henry Jasik, copyright 1961 by McGraw-Hill Book Co., Inc.

AXIAL RATIO OFF AXIS FOR CROSSED DIPOLE

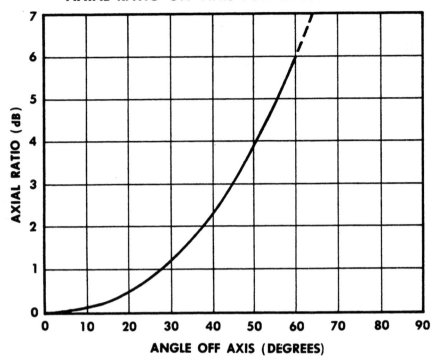

MAXIMUM POLARIZATION LOSS BETWEEN TWO ELLIPTICALLY POLARIZED ANTENNAS

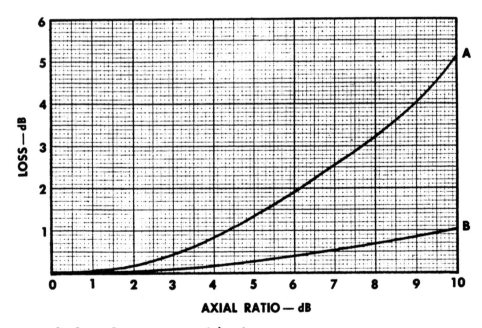

A. Same Sense—same axial ratios

B. Same Sense—one circularly polarized

Top: Reprinted from Antenna Engineering Handbook, edited by Henry Jasik, copyright 1961 by McGraw-Hill Book Co., Inc. Bottom: Courtesy of R. K. Thomas, The Martin Co., Baltimore, Md.

OPTIMALLY MATCHED COMPENSATED BALUN

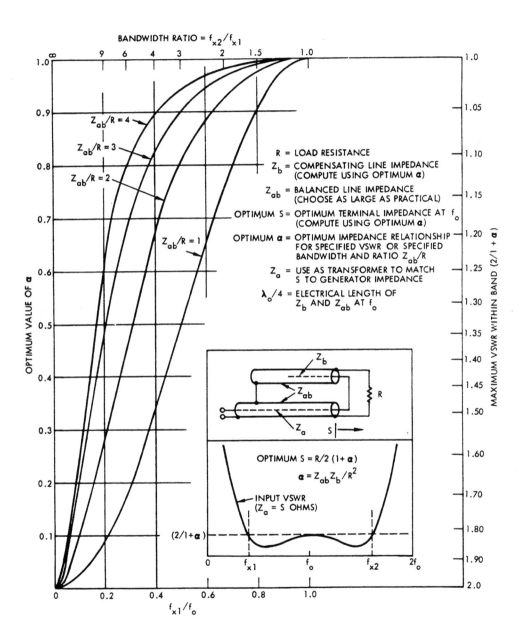

R = LOAD RESISTANCE

Z_b = COMPENSATING LINE IMPEDANCE (COMPUTE USING OPTIMUM α)

Z_{ab} = BALANCED LINE IMPEDANCE (CHOOSE AS LARGE AS PRACTICAL)

OPTIMUM S = OPTIMUM TERMINAL IMPEDANCE AT f_o (COMPUTE USING OPTIMUM α)

OPTIMUM α = OPTIMUM IMPEDANCE RELATIONSHIP FOR SPECIFIED VSWR OR SPECIFIED BANDWIDTH AND RATIO Z_{ab}/R

Z_a = USE AS TRANSFORMER TO MATCH S TO GENERATOR IMPEDANCE

$\lambda_o/4$ = ELECTRICAL LENGTH OF Z_b AND Z_{ab} AT f_o

OPTIMUM S = R/2 (1+ α)

$$\alpha = Z_{ab}Z_b/R^2$$

INPUT VSWR (Z_a = S OHMS)

BANDWIDTH RATIO = f_{x2}/f_{x1}

OPTIMUM VALUE OF α

MAXIMUM VSWR WITHIN BAND (2/1 + α)

$Z_{ab}/R = 4$

$Z_{ab}/R = 3$

$Z_{ab}/R = 2$

$Z_{ab}/R = 1$

51

Courtesy of George Oltman

LONGITUDINAL SHUNT SLOT PARAMETERS

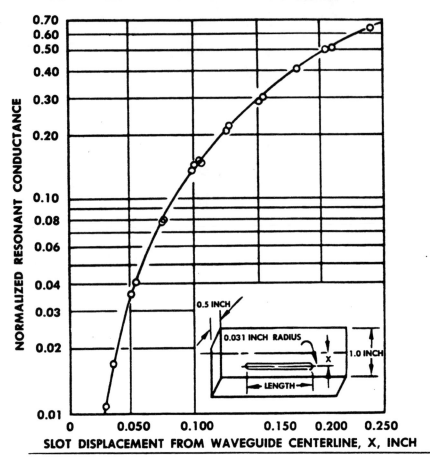

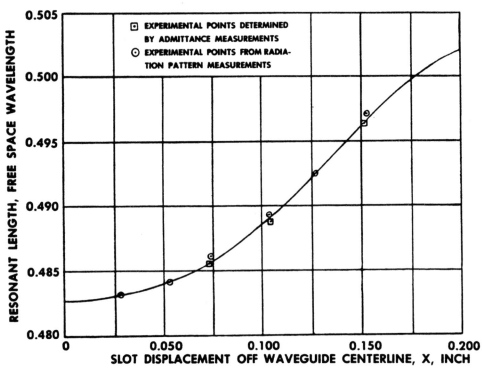

Reprinted from Antenna Engineering Handbook, edited by Henry Jasik, copyright 1961 by McGraw-Hill Book Co., Inc. Courtesy of I. P. Kaminow and R. J. Stegen, formerly of Hughes Aircraft Co.

EDGE SHUNT SLOT PARAMETERS

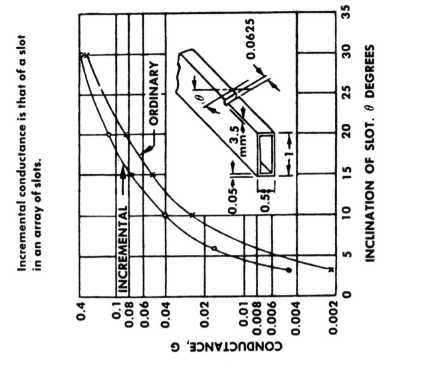

Incremental conductance is that of a slot in an array of slots.

INCREMENTAL AND SINGLE SLOT CONDUCTANCE FOR SHUNT EDGE SLOT (UNMARKED DIMENSIONS OF WAVEGUIDE SECTION ARE IN INCHES) FREQUENCY 9375 MHz

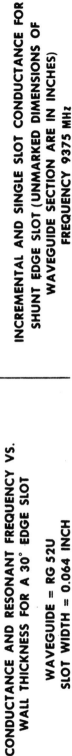

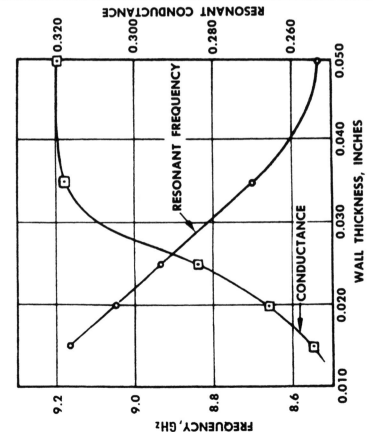

CONDUCTANCE AND RESONANT FREQUENCY VS. WALL THICKNESS FOR A 30° EDGE SLOT

WAVEGUIDE = RG 52U
SLOT WIDTH = 0.064 INCH

Reprinted from Antenna Engineering Handbook, edited by Henry Jasik, copyright 1961 by McGraw-Hill Book Co., Inc. Courtesy of I. P. Kaminow and R. J. Stegen, formerly of Hughes Aircraft Co.

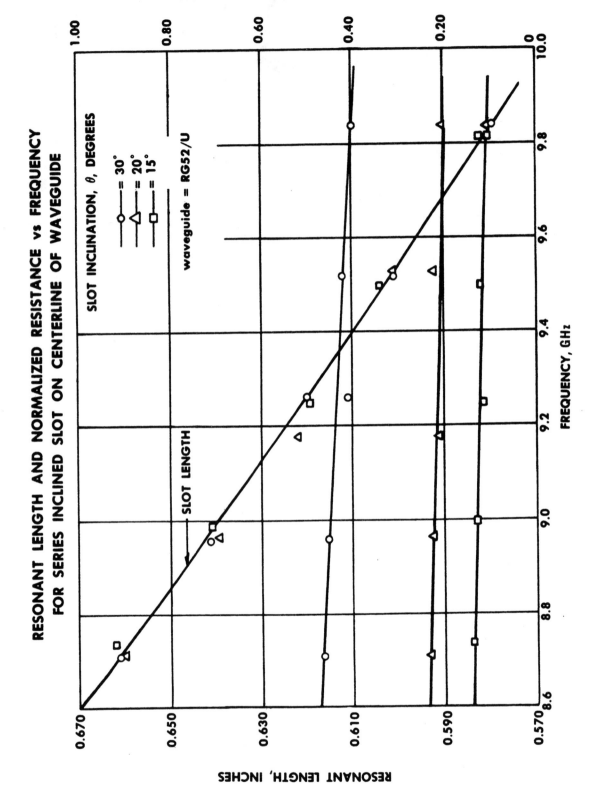

RESONANT LENGTH AND NORMALIZED RESISTANCE vs FREQUENCY
FOR SERIES INCLINED SLOT ON CENTERLINE OF WAVEGUIDE

NORMALIZED RESONANT RESISTANCE

RESONANT LENGTH, INCHES

FREQUENCY, GHz

SLOT INCLINATION, θ, DEGREES

○ = 30°
△ = 20°
□ = 15°

waveguide = RG52/U

SLOT LENGTH

Courtesy of I. P. Kaminow and R. J. Stegen, formerly of Hughes Aircraft Co., Culver City, Cal.

54

DISPLACED AND INCLINED (COMPLEX) SLOT DESIGN

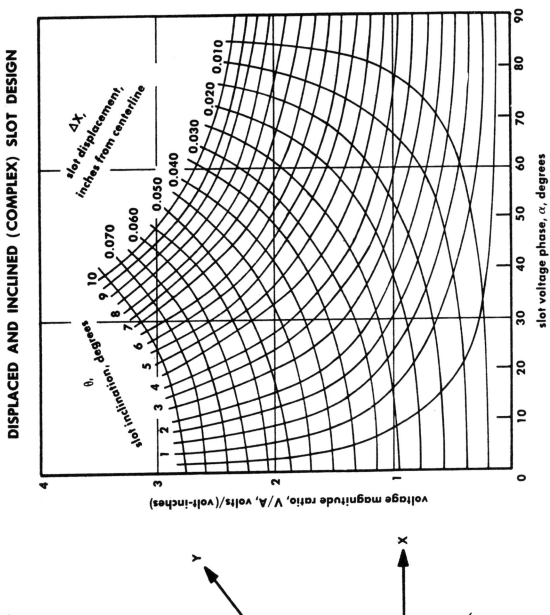

THE SLOT GEOMETRY
(RG-52 waveguide)

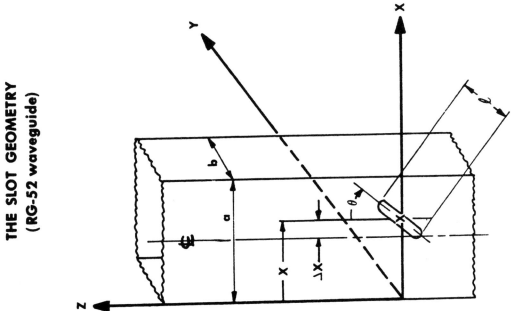

Reprinted from IRE Trans. on Antennas and Propagation, vol. AP-8, no. 4, July 1960; p. 385. Courtesy of B. J. Maxum, formerly with Hughes Aircraft Co.

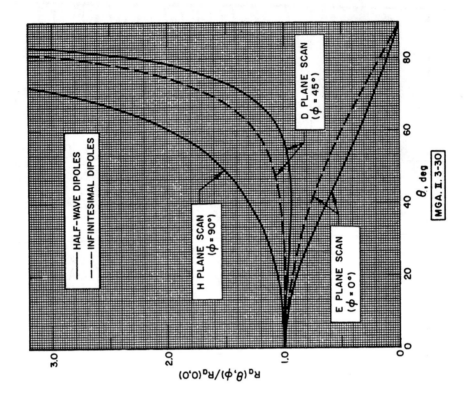

NORMALIZED ACTIVE RESISTANCE VS SCAN ANGLE FOR DIPOLES λ/4 ABOVE A GROUND PLANE (DIPOLES SPACED λ/2 APART)

MGA. II. 3-30

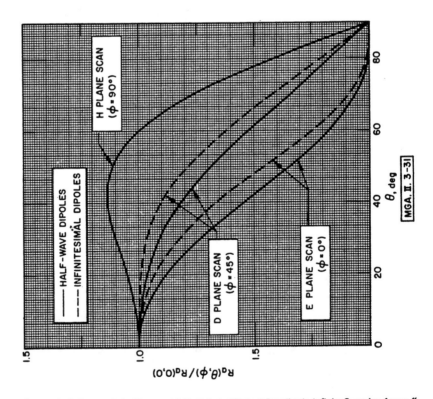

NORMALIZED ACTIVE RESISTANCE VS SCAN ANGLE FOR DIPOLES IN FREE SPACE (DIPOLES SPACED λ/2 APART)

MGA. II. 3-31

Courtesy B.L. Diamond. Reference A.A. Oliner and R.G. Malech, "Mutual Coupling in Infinite Scanning Arrays," Chap. 3 of Vol. II of Microwave Scanning Antennas; *R.C. Hansen, ed., Academic Press, 1966.*

Microwave Engineers'

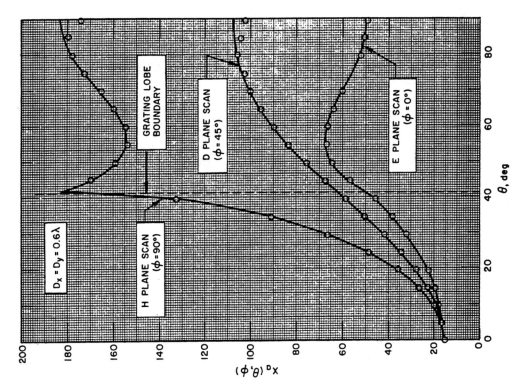

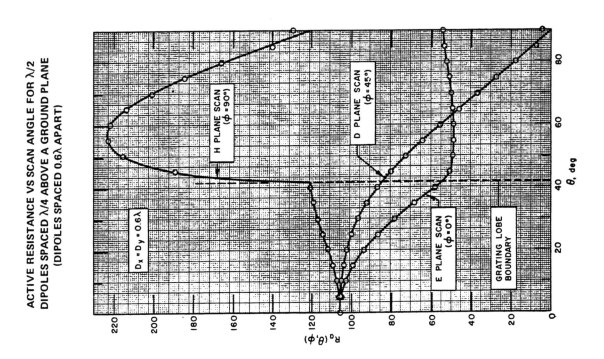

Courtesy B.L. Diamond. Reference A.A. Oliner and R.G. Malech, "Mutual Coupling in Infinite Scanning Arrays," Chap. 3 of Vol. II of Microwave Scanning Antennas; R.C. Hansen, ed., Academic Press 1966.

58

SCANNED LINEAR ARRAY BEAMWIDTH

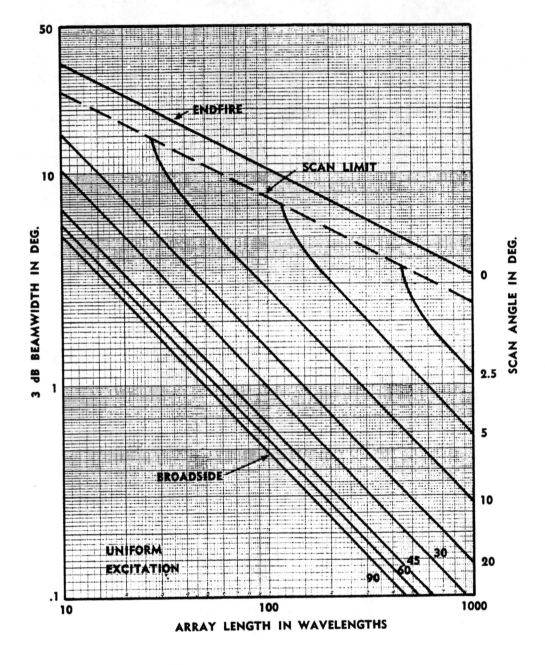

ENDFIRE

SCAN LIMIT

BROADSIDE

UNIFORM EXCITATION

3 dB BEAMWIDTH IN DEG.

ARRAY LENGTH IN WAVELENGTHS

SCAN ANGLE IN DEG.

Reprinted from Microwave Scanning Antennas, edited by R. C. Hansen, Vol. II, Ch. 1, copyright 1965 by Academic Press Inc., New York. Courtesy of R. S. Elliott.

BEAM BROADENING VS. SIDELOBE LEVEL FOR LINEAR ARRAY

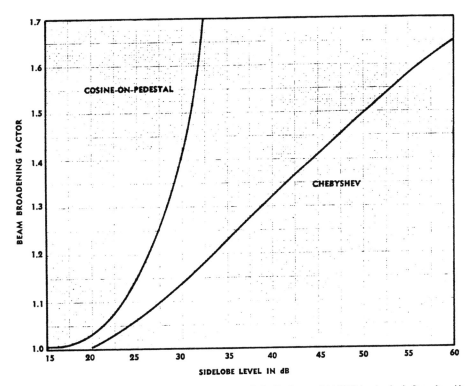

DIRECTIVITY OF UNIFORM SQUARE APERTURE VS. SCAN ANGLE

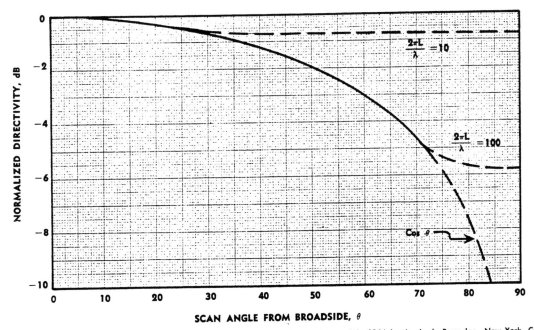

ARRAY DIRECTIVITY VS. ELEMENT SPACING

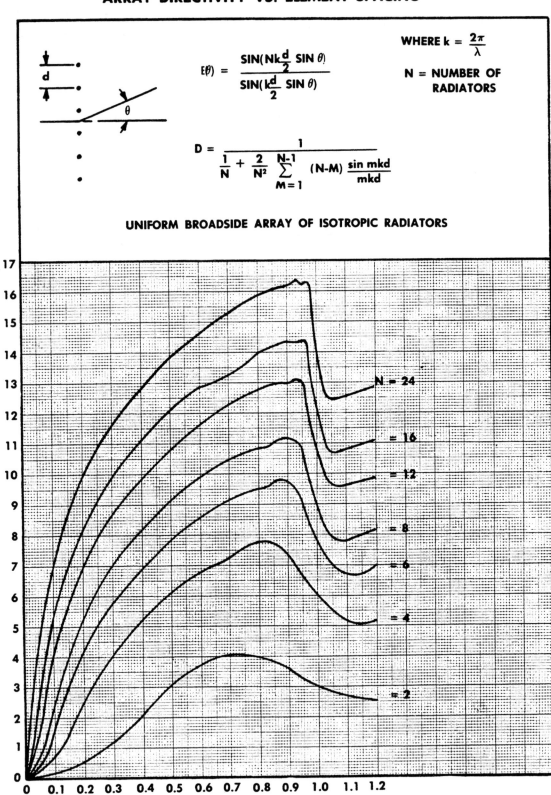

$$E(\theta) = \frac{SIN(Nk\frac{d}{2} SIN\,\theta)}{SIN(k\frac{d}{2} SIN\,\theta)}$$

WHERE $k = \frac{2\pi}{\lambda}$

N = NUMBER OF RADIATORS

$$D = \frac{1}{\frac{1}{N} + \frac{2}{N^2} \sum_{M=1}^{N-1} (N-M)\frac{\sin mkd}{mkd}}$$

UNIFORM BROADSIDE ARRAY OF ISOTROPIC RADIATORS

Courtesy of H. E. King.

SUBTENDED ANGLE AND PATH TAPER OF AN OPTICAL FEED

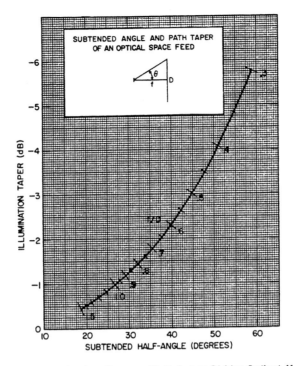

Courtesy B. R. Hatcher, Raytheon Company, Missile Systems Division, Bedford, Massachusetts.

GRATING LOBE POSITION FOR SCANNING LINEAR ARRAY

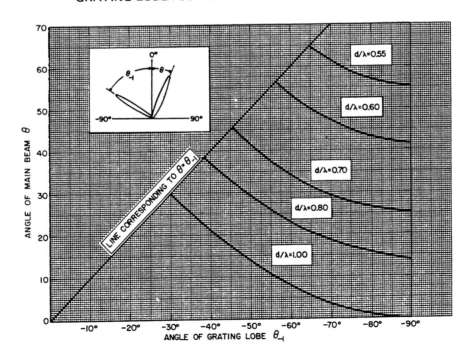

Courtesy W. E. Rupp. Reference A. A. Oliner and R. G. Malech, "Mutual Coupling in Infinite Scanning Arrays," Chap. 3 of Vol. II of "Microwave Scanning Antennas"; R. C. Hansen, ed., Academic Press, New York, 1966.

DIRECTIVITY OF CHEBYSHEV ARRAYS

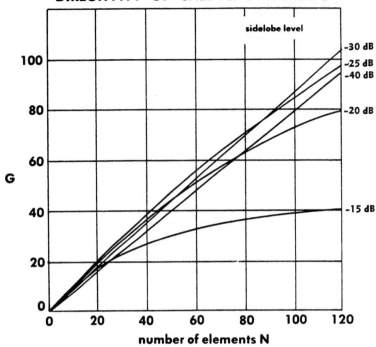

$$G = \frac{2N}{1 + \frac{2}{r^2} \sum_{S=1}^{N-1} \left[T_{2n-1} \left(Z_0 \cos \frac{s\pi}{2N} \right)^2 \right]}$$

$r = T_N(Z_0) = $ sidelobe voltage ratio

BANDWIDTH OF RESONANT ARRAYS

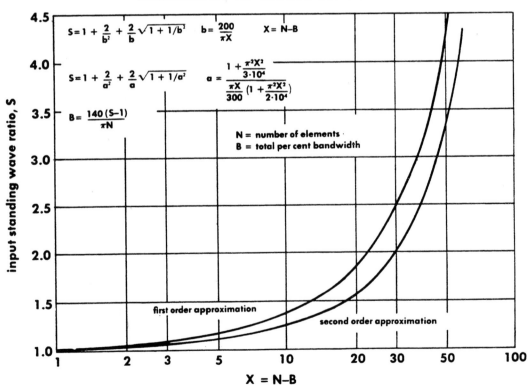

$$S = 1 + \frac{2}{b^2} + \frac{2}{b} \sqrt{1 + 1/b^2} \qquad b = \frac{200}{\pi X} \qquad X = N-B$$

$$S = 1 + \frac{2}{a^2} + \frac{2}{a} \sqrt{1 + 1/a^2} \qquad a = \frac{1 + \frac{\pi^2 X^2}{3 \cdot 10^4}}{\frac{\pi X}{300} \left(1 + \frac{\pi^2 X^2}{2 \cdot 10^4} \right)}$$

$$B = \frac{140(S-1)}{\pi N}$$

N = number of elements
B = total per cent bandwidth

Reprinted from IRE Trans. on Antennas and Propagation, vol. AP-8, Nov., 1960; pp. 629–631. Courtesy of R. J. Stegen and I. P. Kaminow, formerly of Hughes Aircraft Co.

MAXIMUM DIRECTIVITY OF CHEBYSHEV ARRAYS

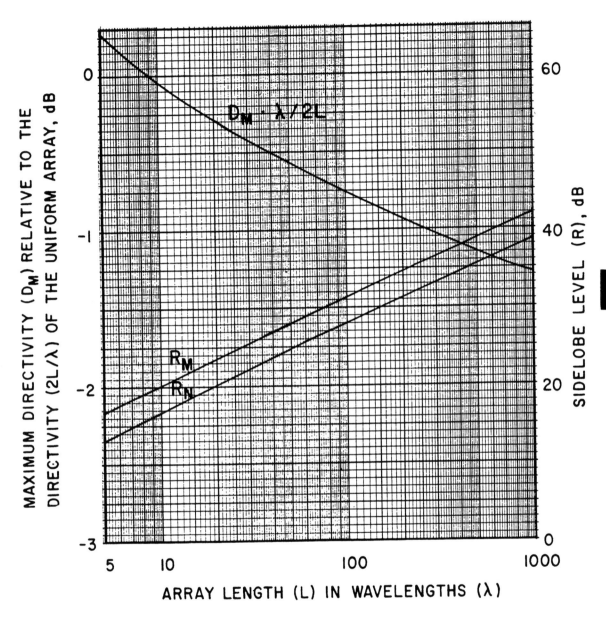

R_M — Sidelobe level for which maximum directivity (D_M) occurs.

R_N — Sidelobe level for which maximum directivity –to–beamwidth ratio occurs.

Courtesy of C. J. Drane

DIRECTIVITY-BEAMWIDTH PRODUCT TO CHEBYSHEV ARRAYS

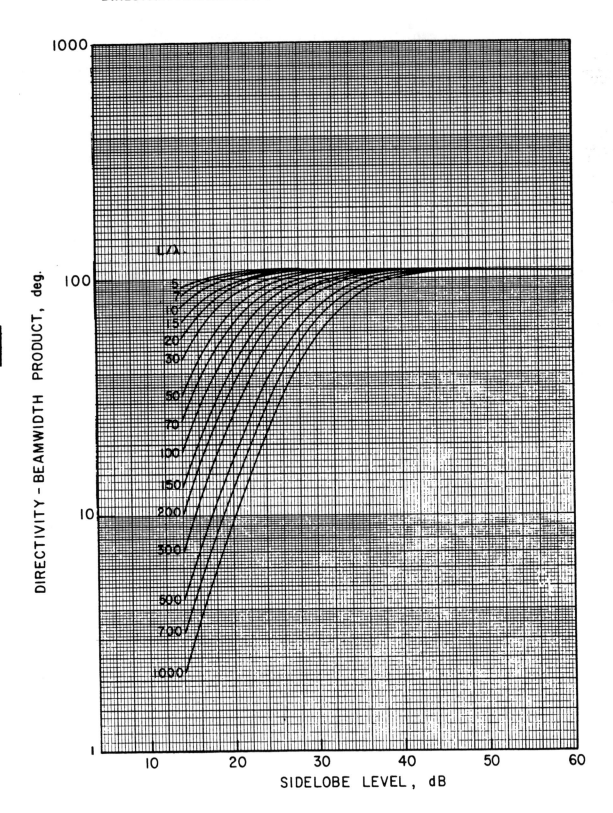

Courtesy of C. J. Drane

DIRECTIVITY-TO-BEAMWIDTH RATIO FOR CHEBYSHEV ARRAYS

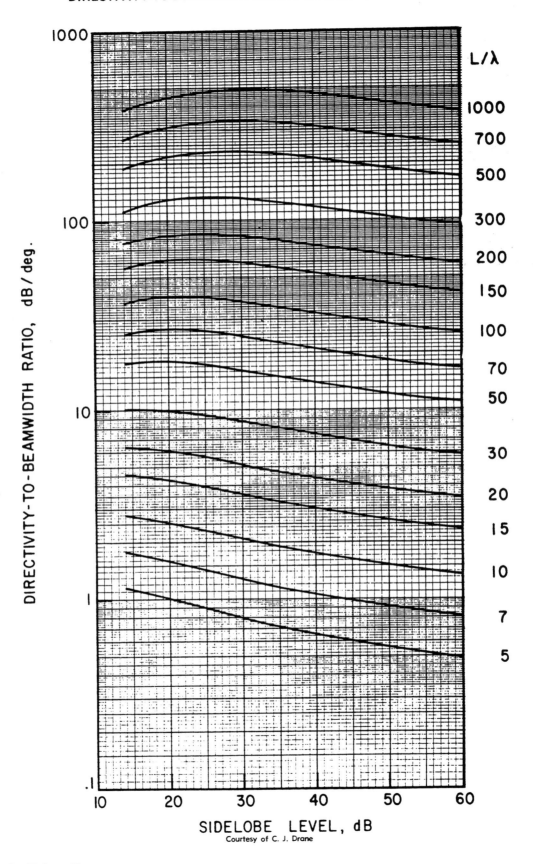

L/λ

Courtesy of C. J. Drane

CONDUCTANCE OF TRAVELING WAVE ARRAY

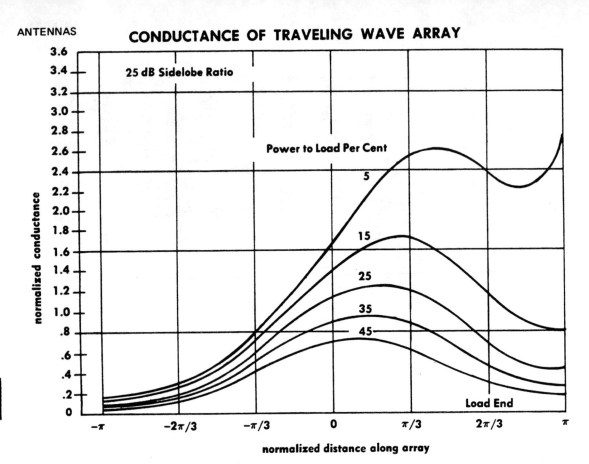

CONDUCTANCE OF TRAVELING WAVE ARRAY

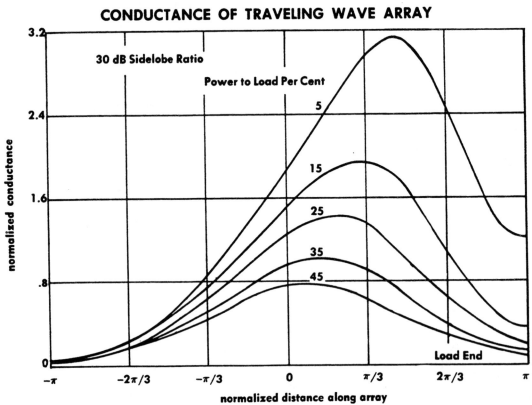

Courtesy of A. Dion, IRE Trans. on Antennas and Propagation, vol. AP-6, Oct., 1958; pp. 360–375.

GAIN VS. NUMBER OF ELEMENTS FOR BROADSIDE SCANNING CHEBISHEV PHASED ARRAY

L = No. of elements one side of square array

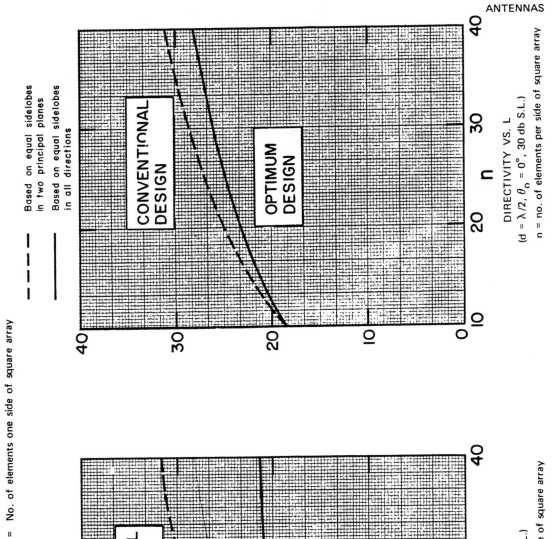

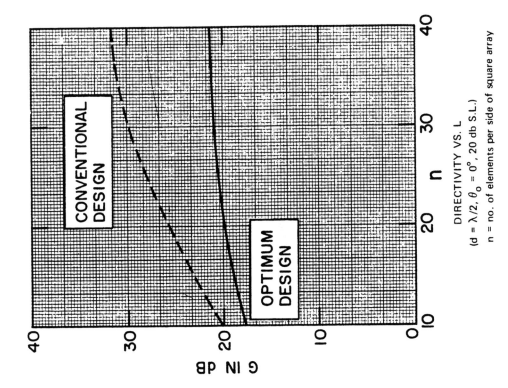

Courtesy D.K. Cheng and F.I. Tseng, Syracuse University. Reference: "Optimum Scannable Planar Arrays with an Invariant Sidelobe Level," Proc. IEEE, Vol. 65, November 1968.

68

MAXIMUM ELEMENT SPACING VS. NUMBER OF ELEMENTS FOR DESIRED GRATING LOBE SUPPRESSION FOR A PHASED ARRAY

d_M/λ = Maximum fractional wavelength spacing

θ_M = Maximum scan angle

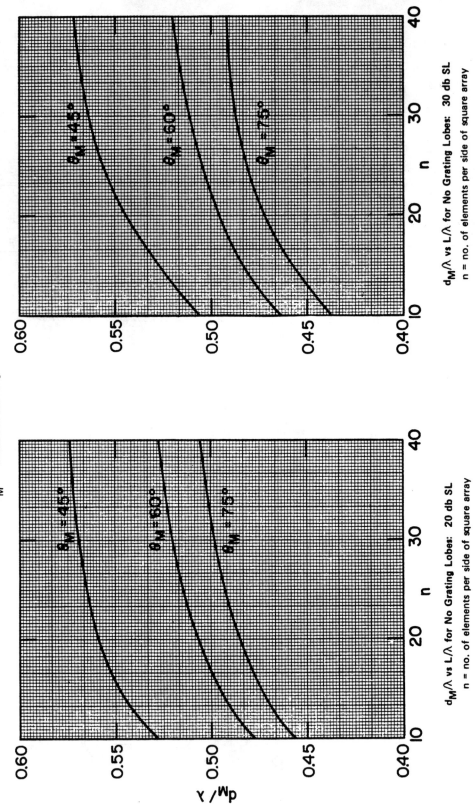

d_M/Λ vs L/Λ for No Grating Lobes: 30 db SL

n = no. of elements per side of square array

d_M/Λ vs L/Λ for No Grating Lobes: 20 db SL

n = no. of elements per side of square array

Courtesy D.K. Cheng and F.I. Tseng, Syracuse University. Reference: "Optimum Scannable Planar Arrays with an Invariant Sidelobe Level," Proc. IEEE, Vol. 56, November 1968.

BEAMWIDTH FOR SCANNING CHEBISHEV PHASED ARRAYS

θ_M maximum scan angle

$(\Delta\theta_C)_M$ 3 db beamwidth

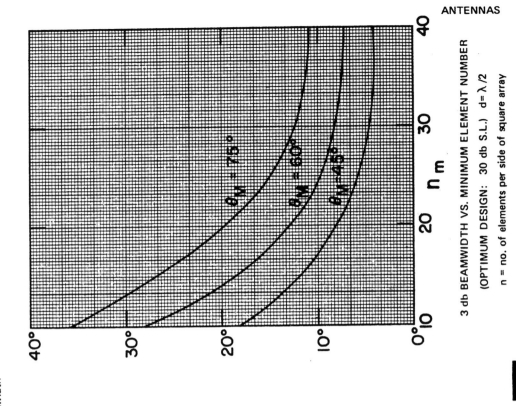

3 db BEAMWIDTH VS. MINIMUM ELEMENT NUMBER

(OPTIMUM DESIGN: 30 db S.L.) $d = \lambda/2$

n = no. of elements per side of square array

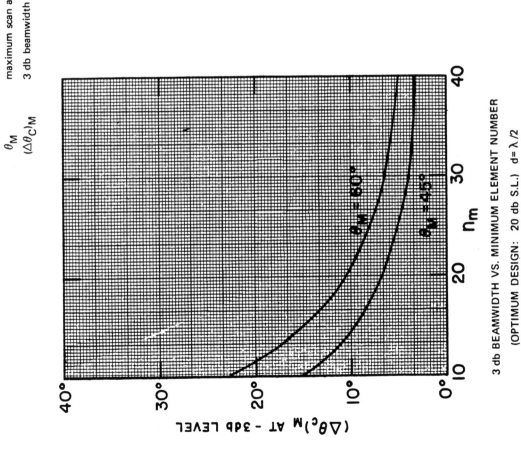

3 db BEAMWIDTH VS. MINIMUM ELEMENT NUMBER

(OPTIMUM DESIGN: 20 db S.L.) $d = \lambda/2$

n = no. of elements per side of square array

Courtesy D.K. Cheng and F.I. Tseng, Syracuse University. Reference: "Optimum Scannable Planar Arrays with an Invariant Sidelobe Level," Proc. IEEE, Vol. 56, November 1968.

69

TAYLOR LINE SOURCE WITH PART EQUAL, PART TAPERED SIDELOBES

β_0 — 3 db beamwidth in radians times L/λ

$\bar{n}$ — Sidelobe where envelope breaks

σ — Beam broadening factor due to $\bar{n}$

TAYLOR LINE SOURCE DESIGN PARAMETERS

Sidelobe level (db)	Sidelobe amplitude ratio	β_0 (rad)	Values of σ							
			$\bar{n}=3$	$\bar{n}=4$	$\bar{n}=5$	$\bar{n}=6$	$\bar{n}=7$	$\bar{n}=8$	$\bar{n}=9$	$\bar{n}=10$
20	10.00	0.893	1.12133	1.10273	1.08701	1.07490				
25	17.78	0.978	1.09241	1.08698	1.07728	1.06834	1.06083	1.05463		
30	31.62	1.057		1.06934	1.06619	1.06079	1.05538	1.05052	1.04628	1.04262
35	56.23	1.131			1.05386	1.05231	1.04923	1.04587	1.04264	1.03970
40	100.0	1.200				1.04298	1.04241	1.04068	1.03858	1.03643

Courtesy T. T. Taylor. Reference: Chap. 1 of Vol. I of Microwave Scanning Antennas, R.C. Hansen, ed., Academic Press, New York, 1964.

70

Large Chebyshev arrays or large Taylor apertures (the Taylor distribution is usually used for a large Chebyshev array) experience a directivity loss when the main beam becomes sufficiently narrow that the power in the sidelobes is comparable. These curves are valid for a line source or for either dimension of a rectangular source.

NORMALIZED DIRECTIVITY VS LENGTH FOR UNIFORM AND IDEAL SPACE FACTORS

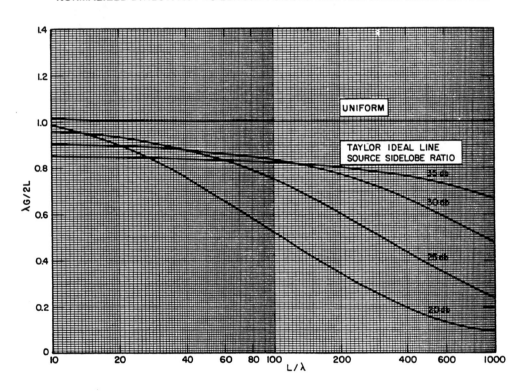

Reference: Chap. 1 of Vol. I of "Microwave Scanning Antennas," R.C. Hansen, ed., Academic Press, New York, 1964. See also: "Gain Limitations of Large Antennas - Correction," R.C. Hansen, IEEE Trans., Vol. AP-13, Nov. 1965, 997-998.

NORMALIZED DIRECTIVITY VS LENGTH FOR TAYLOR APPROXIMATE SPACE FACTOR

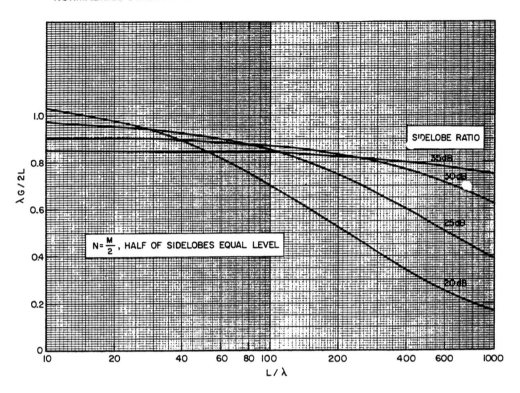

NORMALIZED DIRECTIVITY VS LENGTH FOR TAYLOR APPROXIMATE SPACE FACTOR

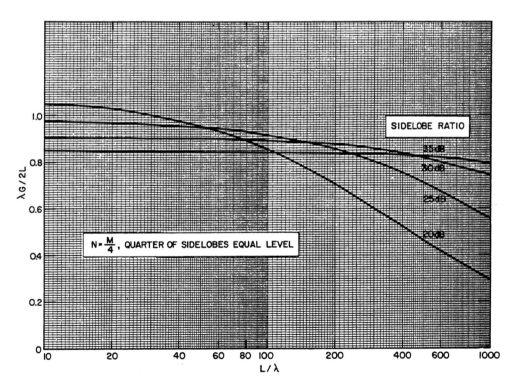

Reference: Chap. 1 of Vol. I of Microwave Scanning Antennas, R.C. Hansen, ed., Academic Press, New York, 1964.
See also: "Gain Limitations of Large Antennas — Correction," R.C. Hansen, IEEE Trans., Vol. AP-13,
Nov. 1965, 997-998.

SIDELOBE LEVEL DEGRADATION FROM CONSTRUCTION ERRORS — SHUNT SLOT ARRAY WITH CHEBYSHEV DISTRIBUTION

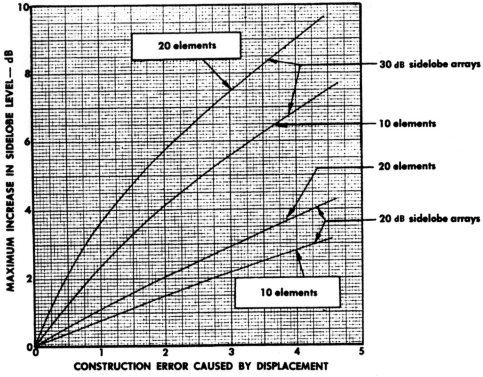

CONSTRUCTION ERROR CAUSED BY DISPLACEMENT OF SLOTS ⊥ TO WAVEGUIDE AXIS-WAVELENGTH × 10³

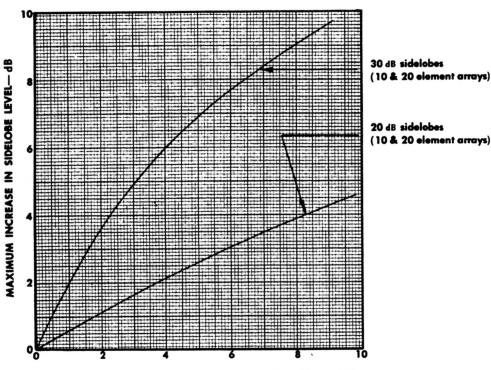

CONSTRUCTION ERROR IN SLOT LENGTH-WAVELENGTH × 10⁴

Courtesy of T. S. Fong, Hughes Aircraft Co., Culver City, Cal.

FIRST BEAM POSITION OFF BROADSIDE, UNIFORM AMPLITUDE, ASYMMETRICAL PHASE

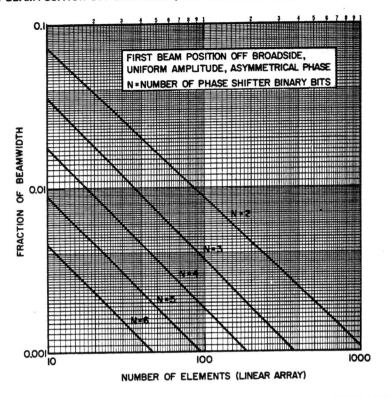

FIRST BEAM POSITION OFF BROADSIDE, UNIFORM AMPLITUDE, SYMMETRICAL PHASE

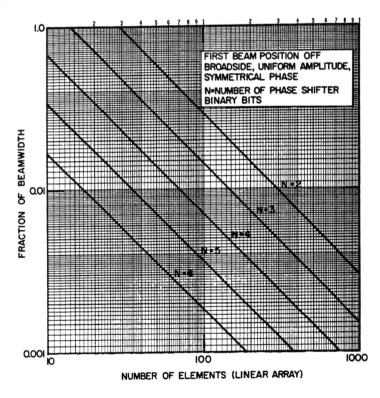

Courtesy B.R. Hatcher, Raytheon Company, Missile Systems Division, Bedford, Massachusetts. To be published in Special Issue of IEEE Proceedings on Electronic Scanning, November 1968.

PATTERN BUILD-UP AS FUNCTION OF TIME, IN 10 ELEMENT ARRAY

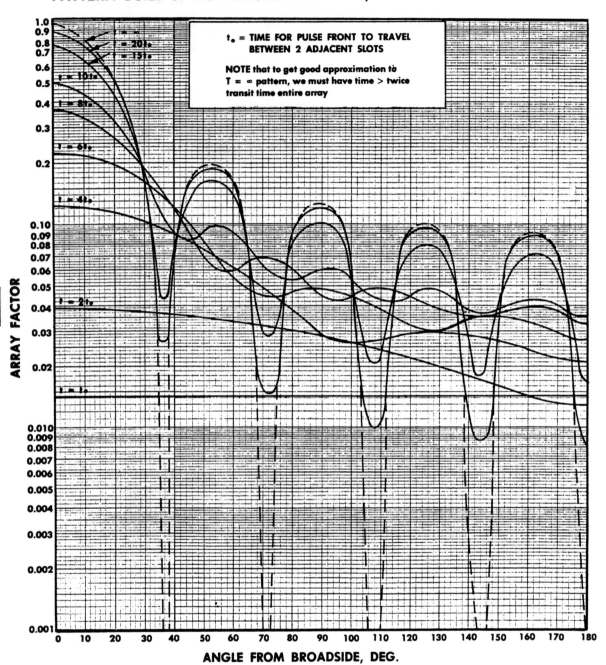

t_o = TIME FOR PULSE FRONT TO TRAVEL BETWEEN 2 ADJACENT SLOTS

NOTE that to get good approximation to $T = \infty$ pattern, we must have time > twice transit time entire array

74

Courtesy of L. L. Bailin, Hughes Aircraft Co., Culver City, Cal.

NON-UNIFORMLY SPACED ARRAYS

Probability of achieving sidelobe level r (main beam = 1) with N elements randomly distributed. Aperture length a.

$$a = 10^q \lambda$$

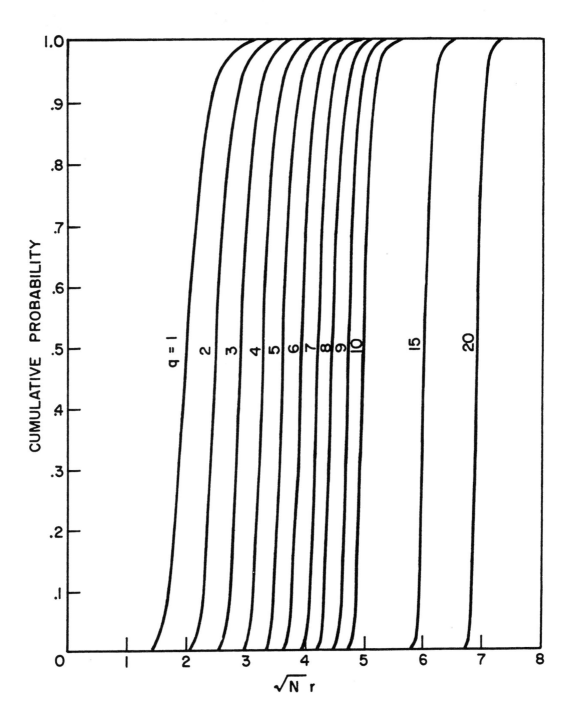

Courtesy Y.T. Lo, University of Illinois Antenna Lab. Reference: Y.T. Lo and R.J. Simcoe, "An Experiment on Antenna Arrays with Randomly Spaced Elements," IEEE Trans., Vol. AP-15, March 1967, 231-235.

76

SYSTEM SENSITIVITY VS. ANTENNA AND RECEIVER NOISE TEMPERATURE

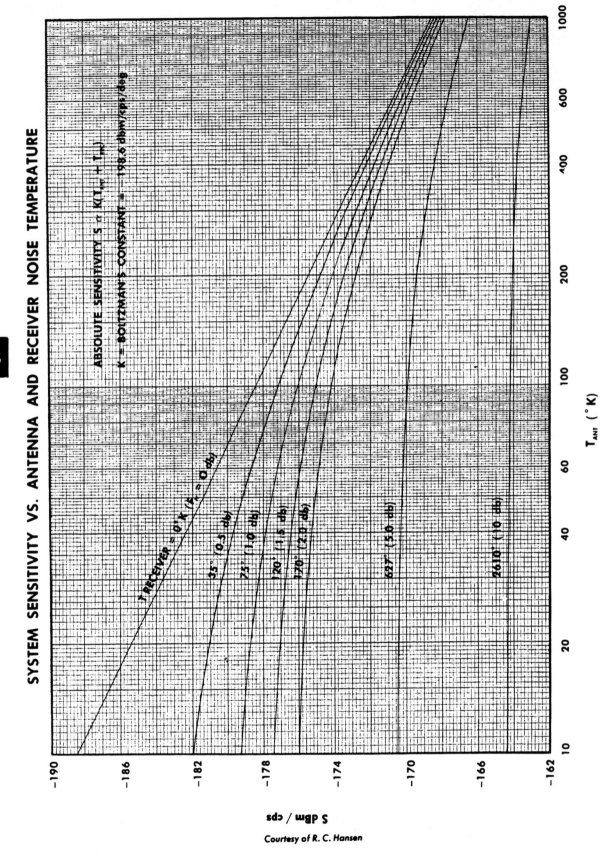

Courtesy of R. C. Hansen

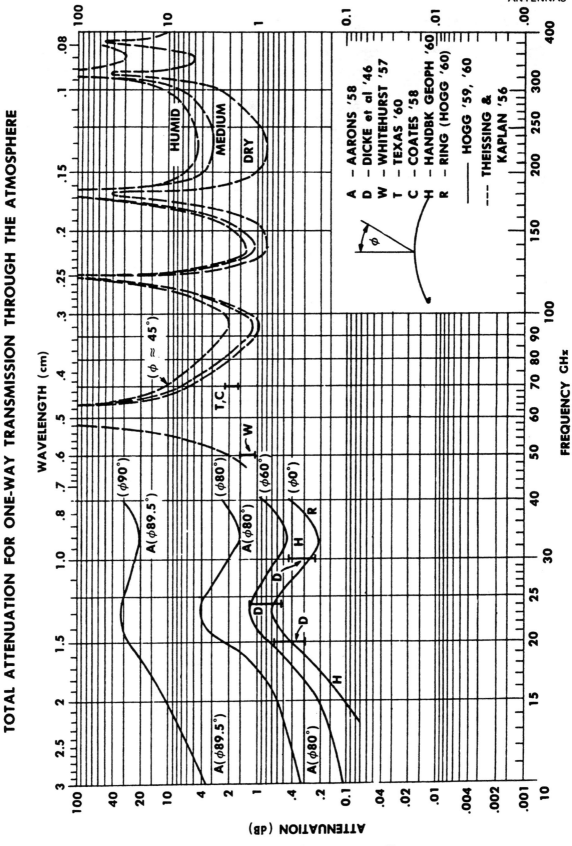

TOTAL ATTENUATION FOR ONE-WAY TRANSMISSION THROUGH THE ATMOSPHERE

ANTENNAS

A — AARONS '58
D — DICKE et al '46
W — WHITEHURST '57
T — TEXAS '60
C — COATES '58
H — HANDBK GEOPH '60
R — RING (HOGG '60)

——— HOGG '59, '60

- - - THEISSING &
KAPLAN '56

77

Courtesy of Lincoln Laboratory, MIT by E. S. Rosenblum. Reprinted by permission.

GALACTIC NOISE TEMPERATURE AFTER KO, BROWN AND HAZARD

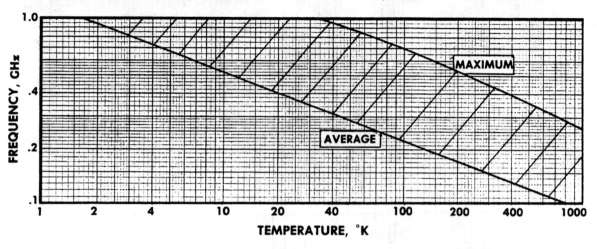

ANTENNA NOISE TEMPERATURE DEGRADATION DUE TO TRANSMISSION LINE LOSS

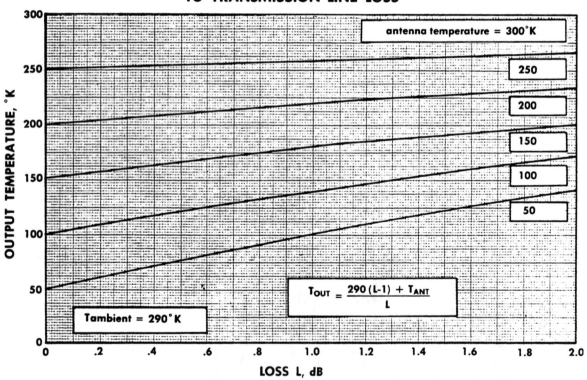

$$T_{OUT} = \frac{290(L-1) + T_{ANT}}{L}$$

Tambient = 290°K

Courtesy of R. C. Hansen

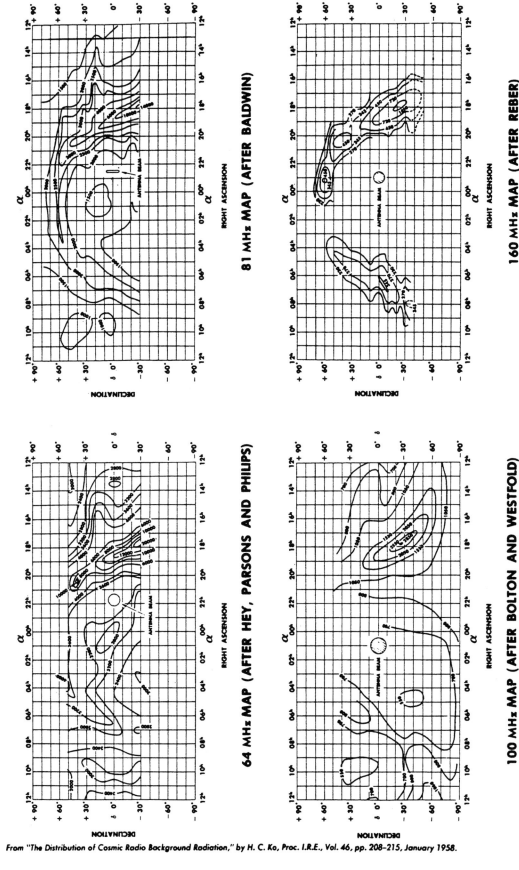

RADIO MAPS OF THE SKY BACKGROUND

THE CONTOURS GIVE THE ABSOLUTE BRIGHTNESS TEMPERATURE OF THE RADIO SKY IN DEGREES KELVIN.

81 MHz MAP (AFTER BALDWIN)

160 MHz MAP (AFTER REBER)

64 MHz MAP (AFTER HEY, PARSONS AND PHILIPS)

100 MHz MAP (AFTER BOLTON AND WESTFOLD)

From "The Distribution of Cosmic Radio Background Radiation," by H. C. Ko, Proc. I.R.E., Vol. 46, pp. 208–215, January 1958.

80

RADIO MAPS OF THE SKY BACKGROUND

THE CONTOURS GIVE THE ABSOLUTE BRIGHTNESS TEMPERATURE OF THE RADIO SKY IN DEGREES KELVIN.

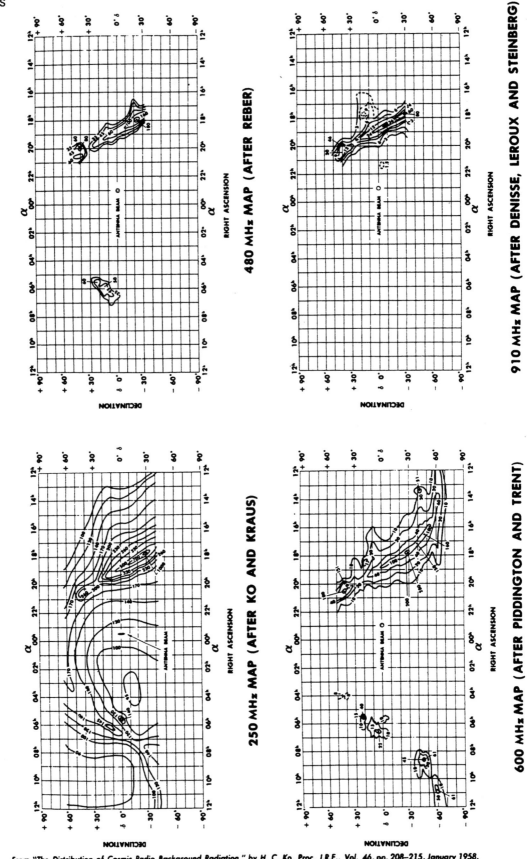

480 MHz MAP (AFTER REBER)

910 MHz MAP (AFTER DENISSE, LEROUX AND STEINBERG)

250 MHz MAP (AFTER KO AND KRAUS)

600 MHz MAP (AFTER PIDDINGTON AND TRENT)

From "The Distribution of Cosmic Radio Background Radiation," by H. C. Ko, Proc. I.R.E., Vol. 46, pp. 208–215, January 1958.

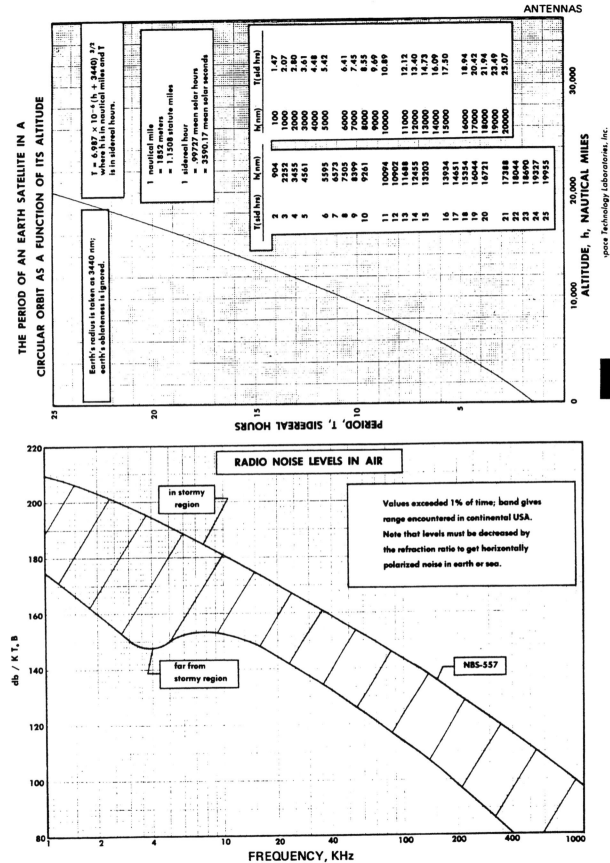

THE PERIOD OF AN EARTH SATELLITE IN A CIRCULAR ORBIT AS A FUNCTION OF ITS ALTITUDE

$T = 6.987 \times 10^{-4} (h + 3440)^{3/2}$ where h is in nautical miles and T is in sidereal hours.

1 nautical mile = 1852 meters = 1.1508 statute miles

1 sidereal hour = .99727 mean solar hours = 3590.17 mean solar seconds

Earth's radius is taken as 3440 nm; earth's oblateness is ignored.

h(nm)	T(sid hrs)
100	1.47
1000	2.07
2000	2.80
3000	3.61
4000	4.48
5000	5.42
6000	6.41
7000	7.45
8000	8.55
9000	9.69
10000	10.89
11000	12.12
12000	13.40
13000	14.73
14000	16.09
15000	17.50
16000	18.94
17000	20.42
18000	21.94
19000	23.49
20000	25.07

T(sid hrs)	h(nm)
2	904
3	2252
4	3455
5	4561
6	5595
7	6573
8	7505
9	8399
10	9261
11	10094
12	10902
13	11688
14	12455
15	13203
16	13934
17	14651
18	15354
19	16044
20	16721
21	17388
22	18044
23	18690
24	19327
25	19955

ALTITUDE, h, NAUTICAL MILES

PERIOD, T, SIDEREAL HOURS

Space Technology Laboratories, Inc.

81

RADIO NOISE LEVELS IN AIR

in stormy region

Values exceeded 1% of time; band gives range encountered in continental USA. Note that levels must be decreased by the refraction ratio to get horizontally polarized noise in earth or sea.

far from stormy region

NBS-557

db / K T, B

FREQUENCY, KHz

Courtesy of R. C. Hansen

ANGLE SUBTENDED BY THE EARTH AT THE SATELLITE AS A FUNCTION OF SATELLITE ALTITUDE

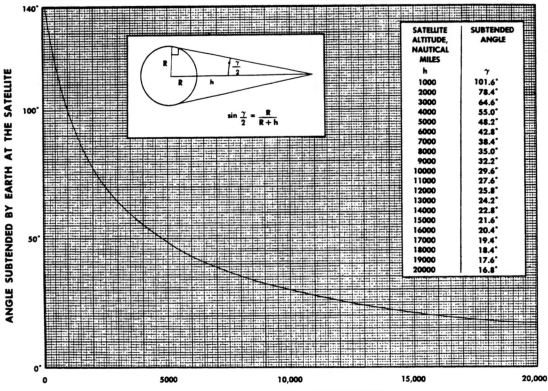

SATELLITE ALTITUDE, NAUTICAL MILES	SUBTENDED ANGLE
h	γ
1000	101.6°
2000	78.4°
3000	64.6°
4000	55.0°
5000	48.2°
6000	42.8°
7000	38.4°
8000	35.0°
9000	32.2°
10000	29.6°
11000	27.6°
12000	25.8°
13000	24.2°
14000	22.8°
15000	21.6°
16000	20.4°
17000	19.4°
18000	18.4°
19000	17.6°
20000	16.8°

$$\sin \frac{\gamma}{2} = \frac{R}{R+h}$$

ANGLE SUBTENDED BY EARTH AT THE SATELLITE (vertical axis)

SATELLITE ALTITUDE, NAUTICAL MILES (horizontal axis)

82

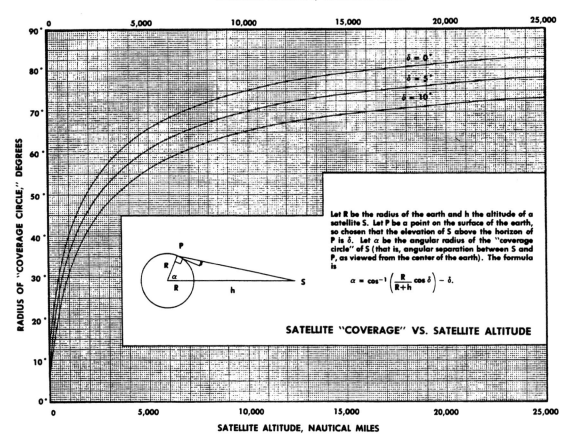

Let R be the radius of the earth and h the altitude of a satellite S. Let P be a point on the surface of the earth, so chosen that the elevation of S above the horizon of P is δ. Let α be the angular radius of the "coverage circle" of S (that is, angular separation between S and P, as viewed from the center of the earth). The formula is

$$\alpha = \cos^{-1}\left(\frac{R}{R+h}\cos\delta\right) - \delta.$$

SATELLITE "COVERAGE" VS. SATELLITE ALTITUDE

RADIUS OF "COVERAGE CIRCLE," DEGREES (vertical axis)

SATELLITE ALTITUDE, NAUTICAL MILES (horizontal axis)

Courtesy of Space Technology Laboratories, Inc.

PLOTTING APPROXIMATE SATELLITE VISIBILITY AREAS USING
ORTHOGRAPHIC MERIDIONAL PROJECTION

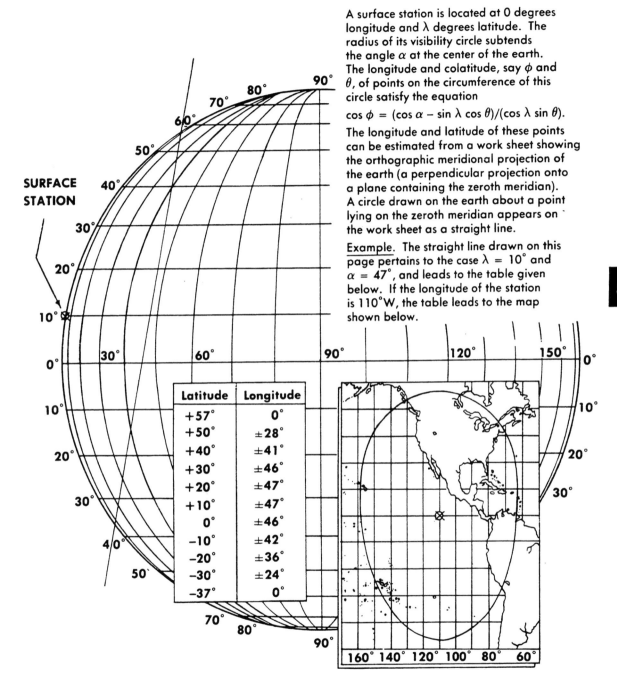

A surface station is located at 0 degrees longitude and λ degrees latitude. The radius of its visibility circle subtends the angle α at the center of the earth. The longitude and colatitude, say ϕ and θ, of points on the circumference of this circle satisfy the equation

$$\cos \phi = (\cos \alpha - \sin \lambda \cos \theta)/(\cos \lambda \sin \theta).$$

The longitude and latitude of these points can be estimated from a work sheet showing the orthographic meridional projection of the earth (a perpendicular projection onto a plane containing the zeroth meridian). A circle drawn on the earth about a point lying on the zeroth meridian appears on the work sheet as a straight line.

Example. The straight line drawn on this page pertains to the case $\lambda = 10°$ and $\alpha = 47°$, and leads to the table given below. If the longitude of the station is 110°W, the table leads to the map shown below.

83

Latitude	Longitude
+57°	0°
+50°	±28°
+40°	±41°
+30°	±46°
+20°	±47°
+10°	±47°
0°	±46°
−10°	±42°
−20°	±36°
−30°	±24°
−37°	0°

Courtesy of Space Technology Laboratories, Inc.

NOMOGRAPH OF PATH LOSS AS A FUNCTION
OF FREQUENCY AND RANGE

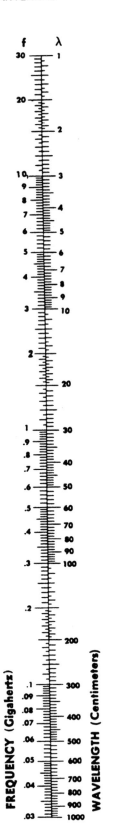

84

f   λ

FREQUENCY (Gigahertz)   **WAVELENGTH (Centimeters)**

LEGEND

$$L_r = L_t + G_t + G_r - N$$

where L_r = Received power level in dBm G_r = Receiving-antenna gain in dB
  L_t = Transmitted power level in dBm λ = Wavelength
  G_t = Transmitting-antenna gain in dB N = $20 \log (4\pi R/\lambda)$

N

PATH LOSS (Decibels)

If the nomograph does not cover the desired ranges, multiply f or R scale by 10N and add 20N dB to path loss scale. N can be positive or negative, and the indicated procedure can be applied successively to the f and R scales.

R

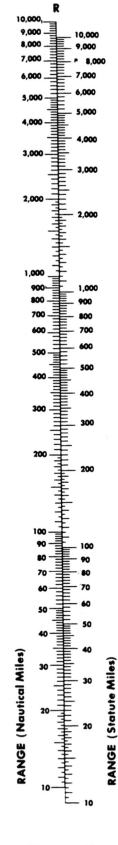

RANGE (Nautical Miles)   **RANGE (Statute Miles)**

TRANSMISSION DELAY TIME

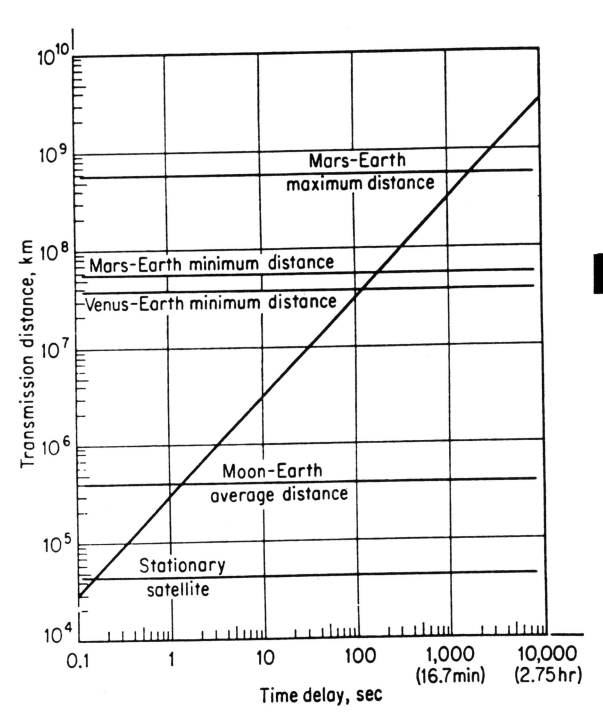

NOMOGRAM, FREE-SPACE TRANSMISSION LOSS, 1000 to 10^6 km

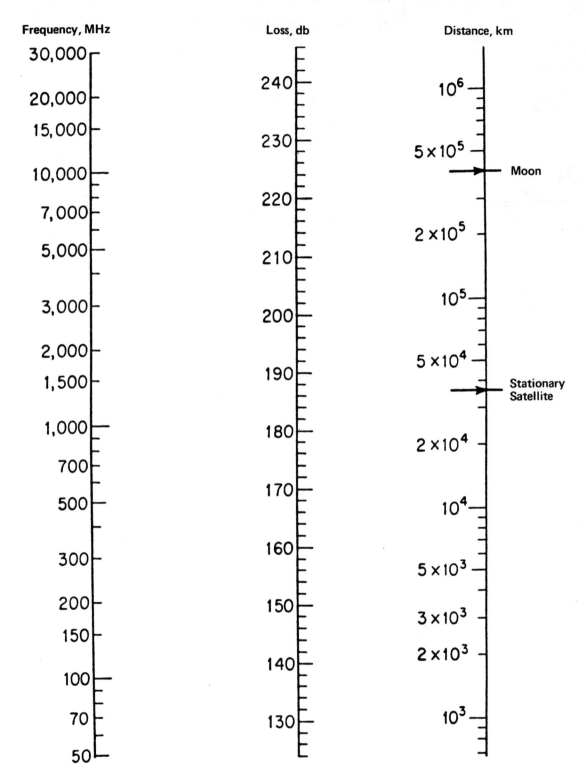

86

NOMOGRAM, FREE-SPACE TRANSMISSION LOSS FOR PLANETARY DISTANCES

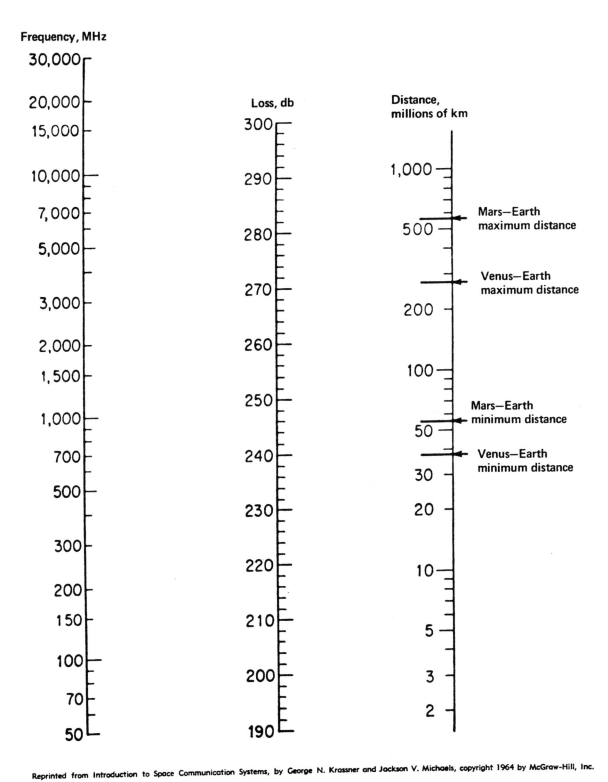

87

Reprinted from Introduction to Space Communication Systems, by George N. Krassner and Jackson V. Michaels, copyright 1964 by McGraw-Hill, Inc.

POINT OF REFLECTION
ON
OVER-WATER MICROWAVE PATH

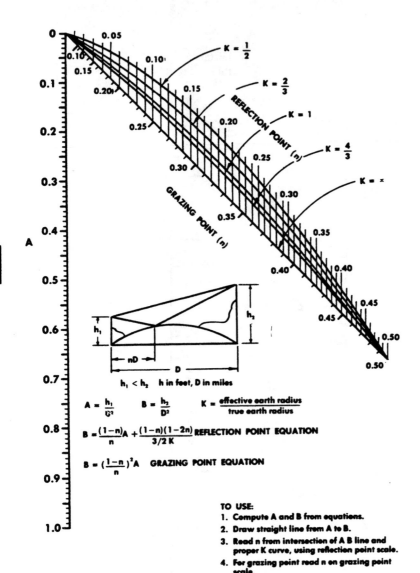

A

B

88

$h_1 < h_2$ h in feet, D in miles

$$A = \frac{h_1}{D^2} \qquad B = \frac{h_2}{D^2} \qquad K = \frac{\text{effective earth radius}}{\text{true earth radius}}$$

$$B = \frac{(1-n)}{n}A + \frac{(1-n)(1-2n)}{3/2\,K} \quad \text{REFLECTION POINT EQUATION}$$

$$B = \left(\frac{1-n}{n}\right)^2 A \quad \text{GRAZING POINT EQUATION}$$

TO USE:
1. Compute A and B from equations.
2. Draw straight line from A to B.
3. Read n from intersection of A B line and proper K curve, using reflection point scale.
4. For grazing point read n on grazing point scale.

QUANTITATIVE EVALUATION AND
COMPARISON OF ELECTRICALLY SMALL ANTENNAS

For the last several years, the antenna section has been devoted to a particular topic. For example, the 1969 Handbook covered arrays and the 1970 Handbook covered aperture antennas. This year, as part of the new Handbook format, R.C. Hansen has provided a set of charts which allow small antennas of all types to be quantatively evaluated and compared.

This set of curves allows a quantitative evaluation of electrically small antennas (largest dimension ≪ λ). As shown by Wheeler, Chu and others, all electrically small antennas are basically dipoles or loops. The first few charts give performance envelopes for dipole antennas. The calculations assume negligible copper loss and no inductive loading. The reactance formula is an empirical best fit by C-T Tai. Bandwidth is given approximately by 1/Q. The loss introduced by an impedance matching coil at the terminals is assumed to involve a coil Q=500, as this is a practical upper bound. The important parameters for evaluating transmitting performance are efficiency and efficiency x bandwidth. The important parameters for evaluating receiving performance are effective length (normalized by λ) and tuning circuit voltage per unit incident electric field, again normalized. An improvement in radiation resistance by a factor of 2.8 can be obtained at a single frequency through inductive loading. Other types of loading and semiconductor devices offer no significant improvement unless located at the dipole terminals.

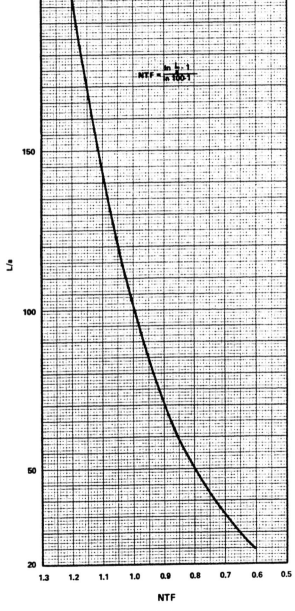

Courtesy of R. C. Hansen

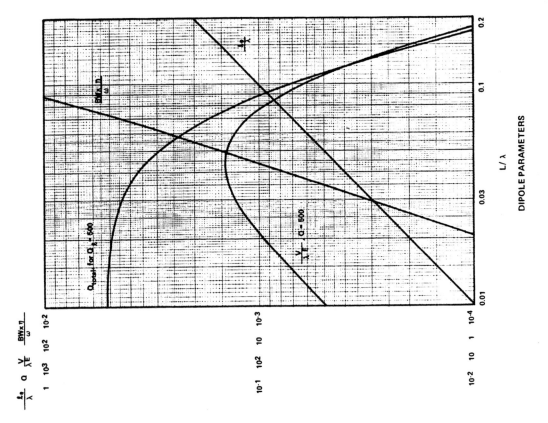

Courtesy of R. C. Hansen

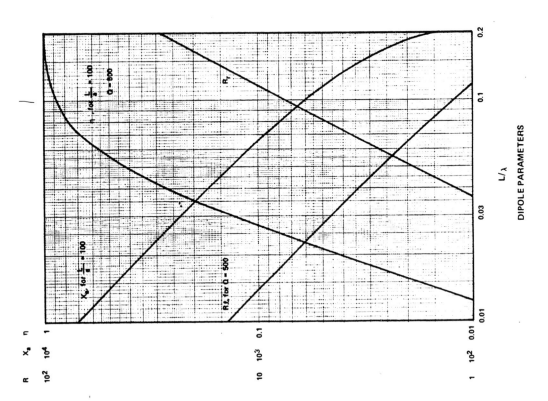

The following curves allow quantitative evaluation of electrically small multi-turn air core loops. The calculations are for multilayer solenoids of a square cross section as this cross section is close to that which gives optimum Q and requires only one dimension to be specified. The wire diameter is assumed to be >> skin depth. In the development of the formulas the convenient loss factor is found to be $R_s D/a$ where R_s is the surface resistivity and D/a is the loop diameter to wire radius. This parameter is likely to be somewhat invariant as larger diameter loops are usually made of larger diameter wire. The merit factors for loops are the same as those for dipoles. Single-turn air core loops with large conductors have also been evaluated but curves are not shown here. The loss and reactance formulas are somewhat different for single turn. Depending on the specific design, a single-turn air loop may be comparable to a multi-turn air loop for transmitting but the single-turn loop is generally inferior for receiving. Loops are usually acceptable only for receiving applications.

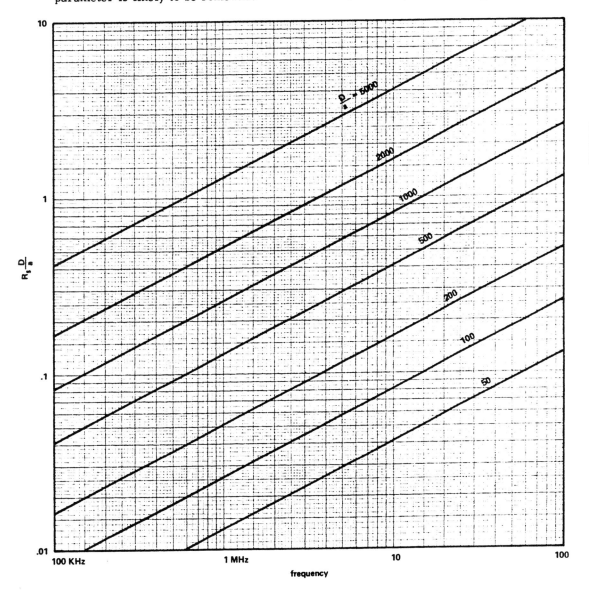

LOSS FACTOR FOR COPPER

Courtesy of R. C. Hansen

92

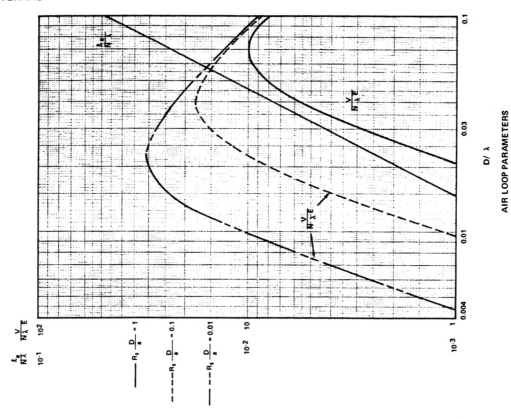

AIR LOOP PARAMETERS

D/λ

Courtesy of R. C. Hansen

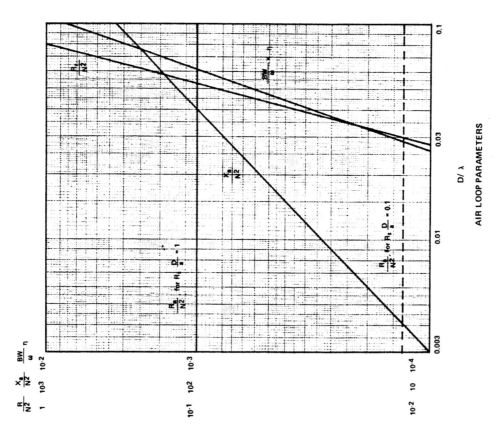

AIR LOOP PARAMETERS

D/λ

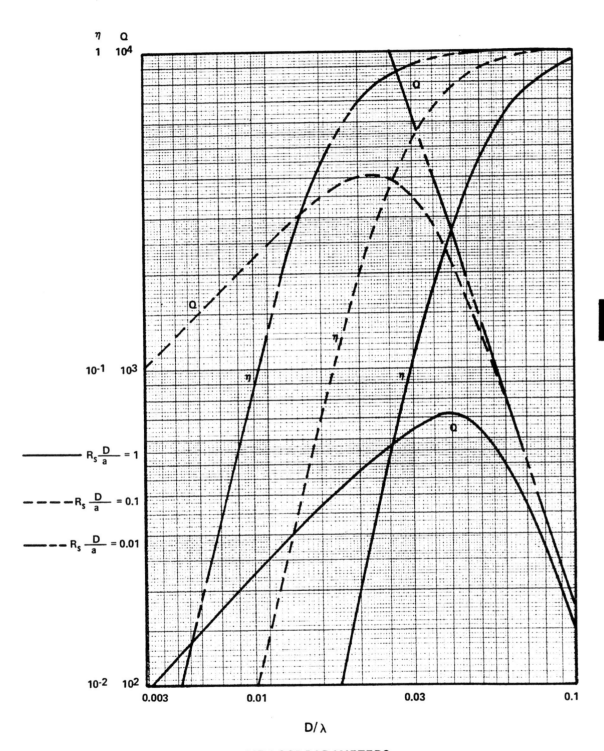

AIR LOOP PARAMETERS

Courtesy of R. C. Hansen

94

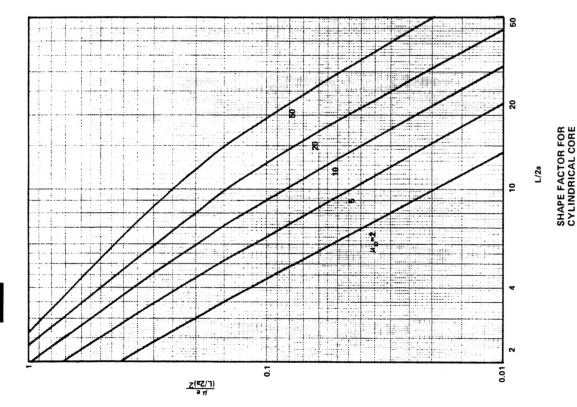

SHAPE FACTOR FOR CYLINDRICAL CORE

L/2a

$$\frac{\mu_e}{(L/2a)^2}$$

Courtesy of R. C. Hansen

The following curves allow quantitative evaluation of electrically small ferrite loop antennas. To realize high effective permeability, the length-to-diameter ratio of the ferrite core should be large. The core need not be round in cross section as long as the length is $\gg$ the largest cross-section dimension. Considerable weight saving can occur through use of hollow cores at a slight degradation in performance. All data here, however, are for solid cores. It has been assumed that the core and coil are circular in cross section and in the formulas the effective permeability enters conveniently as the shape factor $\mu_e/(L/2a)^2$ where $L/2a$ is the ratio of core length to coil diameter. The Q has been assumed to be constant and equal to 500 as this is a reasonable upper limit for Ferramic ferrites in the HF frequency range. The merit factors for evaluating performance are the same as those used for dipole antennas. The performance of ferrite loops and air core loops can be compared using these curves and those immediately preceeding. Again loops are usually acceptable only for receiving applications.

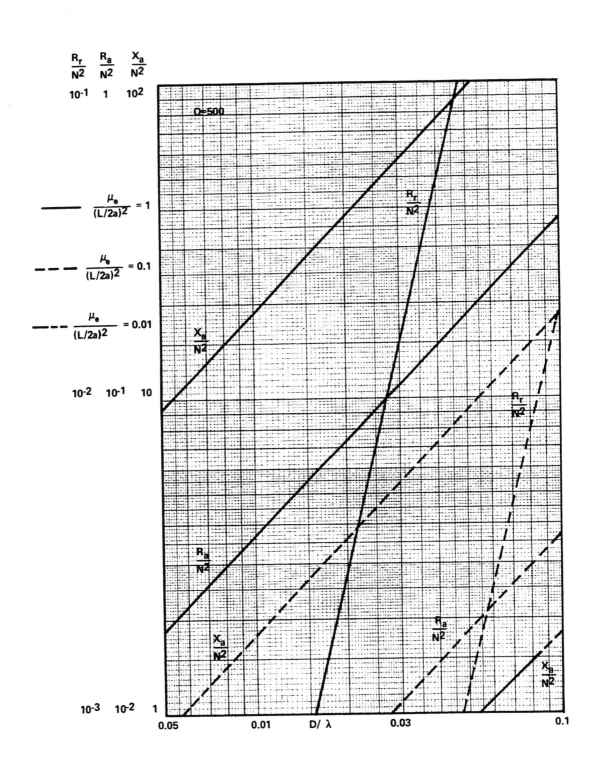

FERRITE LOOP PARAMETERS

Courtesy of R. C. Hansen

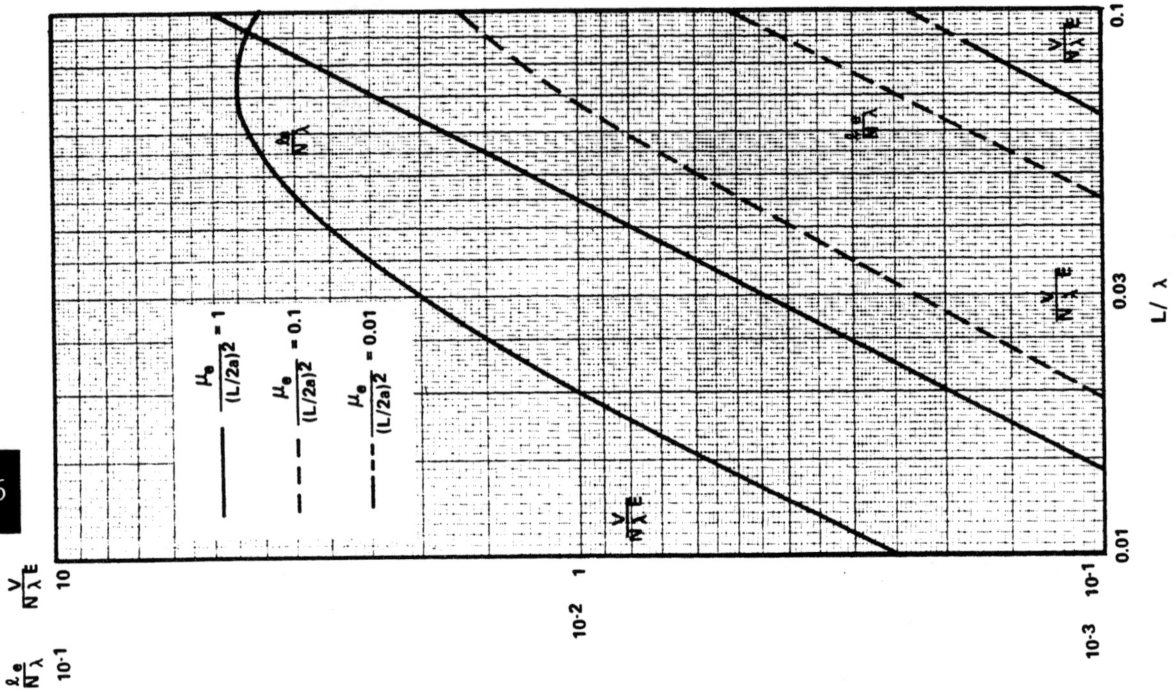
$\frac{V}{N\lambda E}$
$\frac{\ell_e}{N\lambda}$
10
1
10^{-1}
10^{-2}
10^{-3}
0.01
0.03
0.1
L/ λ
$\frac{\mu_e}{(L/2a)^2} = 1$
$\frac{\mu_e}{(L/2a)^2} = 0.1$
$\frac{\mu_e}{(L/2a)^2} = 0.01$

FERRITE LOOP PARAMETERS

Courtesy of R. C. Hansen

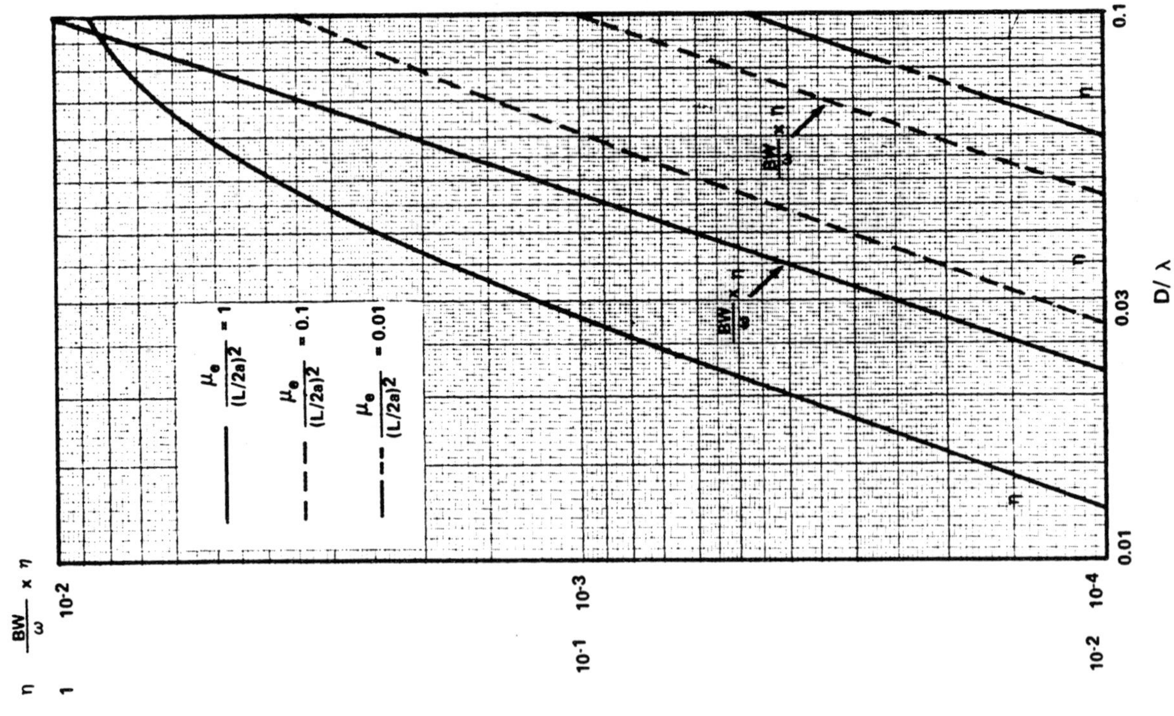
η
$\frac{BW}{\omega} \times \eta$
1
10^{-1}
10^{-2}
10^{-3}
10^{-4}
0.01
0.03
0.1
D/ λ
$\frac{\mu_e}{(L/2a)^2} = 1$
$\frac{\mu_e}{(L/2a)^2} = 0.1$
$\frac{\mu_e}{(L/2a)^2} = 0.01$

FERRITE LOOP PARAMETERS

Quasi-Optical Apertures

TYPE	RAY DIAGRAM	OPTICAL ELEMENTS	PERTINENT DESIGN CHARACTERISTICS
PARABOLOID		Reflective M_p = Paraboloidal mirror	1. Free from spherical aberration. 2. Suffers from off-axis coma. 3. Available in small and large diameters and f/numbers. 4. Low IR loss (Reflective). 5. Detector must be located in front of optics.
CASSEGRAIN		Reflective M_p = Paraboloidal mirror M_s = Hyperboleidel mirror	1. Free from spherical aberration. 2. Shorter than Gregorian. 3. Permits location of detector behind optical system. 4. Quite extensively used.
GREGORIAN		Reflective M_p = Paraboloidal mirror M_s = Ellipsoidal mirror	1. Free from spherical aberration. 2. Longer than cassegrain. 3. Permits location of detector behind optical system. 4. Gregorian less common than cassegrain.
NEWTONIAN		Reflective M_p = Paraboloidal mirror M_s = Reflecting prism or plane mirror	1. Suffers from off-axis coma. 2. Central obstruction by prism or mirror.
HERSCHELIAN		Reflective M_p = Paraboloidal mirror inclined axis	1. Not widely used now. 2. No central obstruction by auxiliary lens. 3. Simple construction. 4. Suffers from some coma.
FRESNEL LENS		Refractive L_p = Special fresnel lens	1. Free of spherical aberration. 2. Inherently lighter weight. 3. Small axial space. 4. Small thickness reduced infrared absorption. 5. Difficult to produce with present infrared transmitting materials.
MANGIN MIRROR		Refractive-reflective M_p = Spherical refractor M_s = Spherical reflector	1. Suitable for IR Source systems. 2. Free of spherical aberration. 3. Most suitable for small apertures. 4. Covers small angular field. 5. Uses spherical surfaces.

Courtesy of George F. Levy. Reference: "Infra-Red System Design," EDN, Vol. pp.   May 1958.

FERRITES

NORMALIZED INTERNAL DC FIELD AT RESONANCE AS A FUNCTION OF SHAPE FOR SPHEROIDAL SAMPLES

$$K = \frac{\omega/\gamma}{4\pi M_s}$$

NORMALIZED EXTERNAL FIELD FOR RESONANCE AS A FUNCTION OF SHAPE FOR SPHEROIDAL SAMPLES.

$$K = \frac{\omega/\gamma}{4\pi M_s}$$

P. Hlawiczka and A. R. Mortis, "Gyromagnetic Resonance Graphical Design Data," Proceedings I. E. E., Vol. 110, No. 4, April 1963.

NORMALIZED EXTERNAL FIELD FOR A VERY SLENDER ELLIPSOID

The dotted line indicates resonant shapes insensitive to temperature variations.

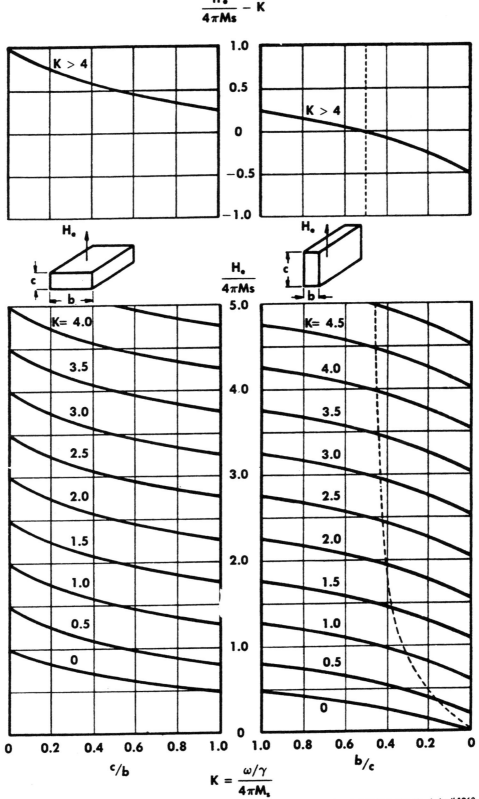

$$\frac{H_e}{4\pi Ms} - K$$

$$K = \frac{\omega/\gamma}{4\pi M_s}$$

P. Hlawiczka and A. R. Mortis, "Gyromagnetic Resonance Graphical Design Data," Proceedings I. E. E., Vol. 110, No. 4, April 1963.

NORMALIZED INTERNAL FIELD FOR A VERY SLENDER ELLIPSOID

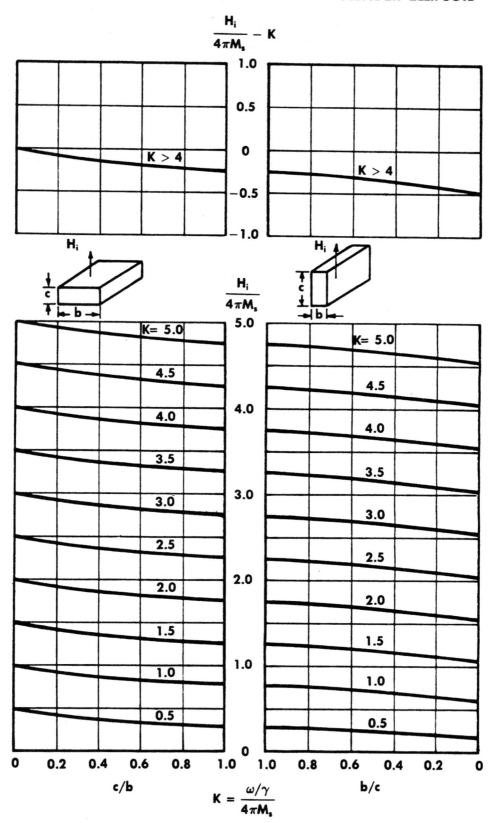

$$\frac{H_i}{4\pi M_s} - K$$

$$\frac{H_i}{4\pi M_s}$$

$$K = \frac{\omega/\gamma}{4\pi M_s}$$

P. Hlawiczka and A. R. Mortis, "Gyromagnetic Resonance Graphical Design Data," Proceedings I. E. E., Vol. 110, No. 4, April 1963.

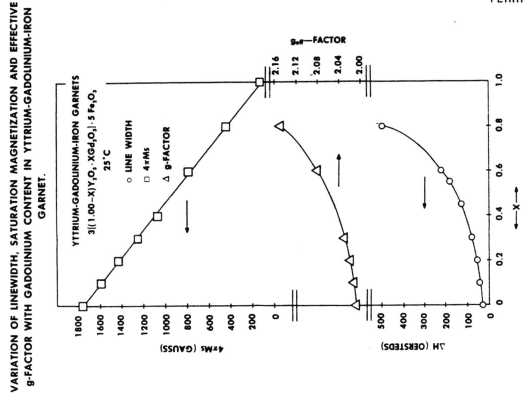

VARIATION OF LINEWIDTH, SATURATION MAGNETIZATION AND EFFECTIVE g-FACTOR WITH GADOLINIUM CONTENT IN YTTRIUM-GADOLINIUM-IRON GARNET.

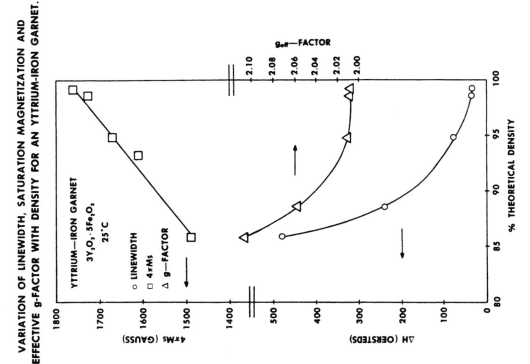

VARIATION OF LINEWIDTH, SATURATION MAGNETIZATION AND EFFECTIVE g-FACTOR WITH DENSITY FOR AN YTTRIUM-IRON GARNET.

Reprinted from the microwave journal, vol. 4, No. 6, 53 (June, 1961). Courtesy of Gordon R. Harrison and L. R. Hodges, Jr., Sperry Microwave Electronics Co., Clearwater, Fla.

103

VARIATION OF SATURATION MAGNETIZATION WITH TEMPERATURE FOR MIXED YTTRIUM-GADOLINIUM-IRON GARNET

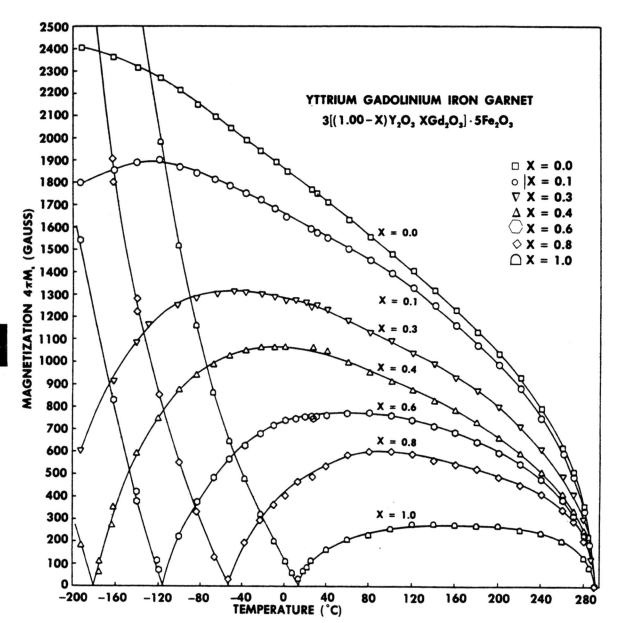

YTTRIUM GADOLINIUM IRON GARNET

$$3[(1.00-X)Y_2O_3, XGd_2O_3] \cdot 5Fe_2O_3$$

□ X = 0.0
○ |X = 0.1
▽ X = 0.3
△ X = 0.4
⬡ X = 0.6
◇ X = 0.8
⌂ X = 1.0

Courtesy of Gordon R. Harrison and L. R. Hodges, Jr., Sperry Microwave Electronics Co., Clearwater, Fla.

VARIATION OF LINEWIDTH, SATURATION MAGNETIZATION AND EFFECTIVE g-FACTOR WITH SAMARIUM CONTENT IN YTTRIUM-SAMARIUM-IRON GARNET.

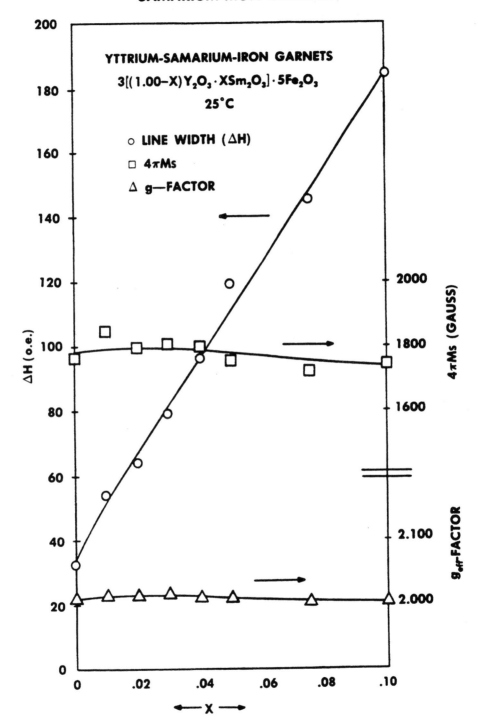

YTTRIUM-SAMARIUM-IRON GARNETS

$3[(1.00-X)Y_2O_3 \cdot XSm_2O_3] \cdot 5Fe_2O_3$

25°C

○ LINE WIDTH (ΔH)
□ $4\pi Ms$
△ g—FACTOR

105

Reprinted from the microwave journal, vol. 4, No. 6, 53 (June 1961). Courtesy of Gordon R. Harrison and L. R. Hodges, Jr., Sperry Microwave Electron: Co., Clearwater, Fla.

VARIATION OF LINEWIDTH, SATURATION MAGNETIZATION, EFFECTIVE g-FACTOR, AND CURIE TEMPERATURE WITH AN ALUMINUM CONTENT IN YTTRIUM-ALUMINUM-IRON GARNET.

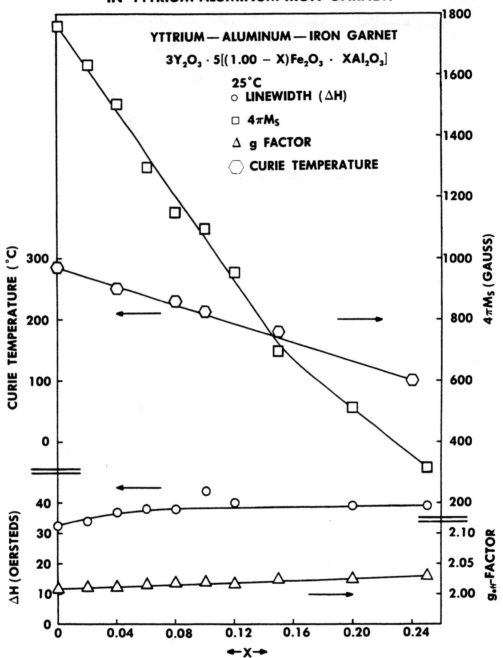

Reprinted from the microwave journal, vol. 4, No. 6, 53 (June, 1961). Courtesy of Gordon R. Harrison and L. R. Hodges, Jr., Sperry Microwave Electronics Co., Clearwater, Fla.

PROPERTIES OF FERRITE MATERIALS

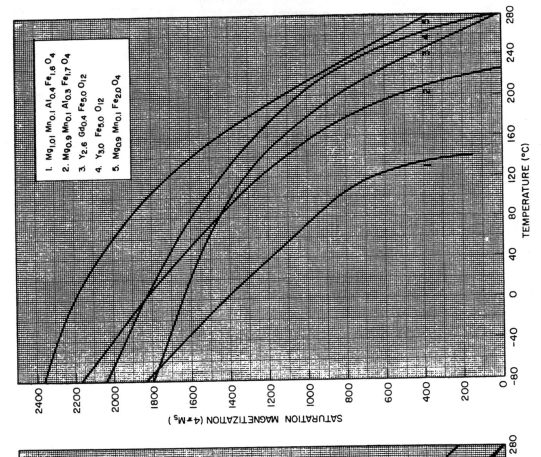

Legend (both graphs):
1. $Mg_{1.01}Mn_{0.1}Al_{0.4}Fe_{1.6}O_4$
2. $Mg_{0.9}Mn_{0.1}Al_{0.3}Fe_{1.7}O_4$
3. $Y_{2.6}Gd_{0.4}Fe_{5.0}O_{12}$
4. $Y_{3.0}Fe_{5.0}O_{12}$
5. $Mg_{0.9}Mn_{0.1}Fe_{2.0}O_4$

Upper graph: SATURATION MAGNETIZATION $(4\pi M_s)$ vs TEMPERATURE (°C)

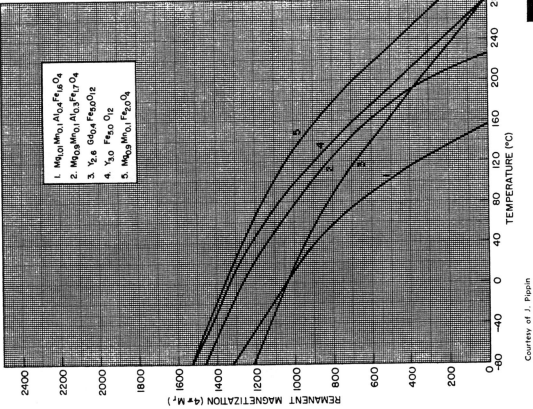

Lower graph: REMANENT MAGNETIZATION $(4\pi M_r)$ vs TEMPERATURE (°C)

Courtesy of J. Pippin

PROPERTIES OF FERRITE MATERIALS

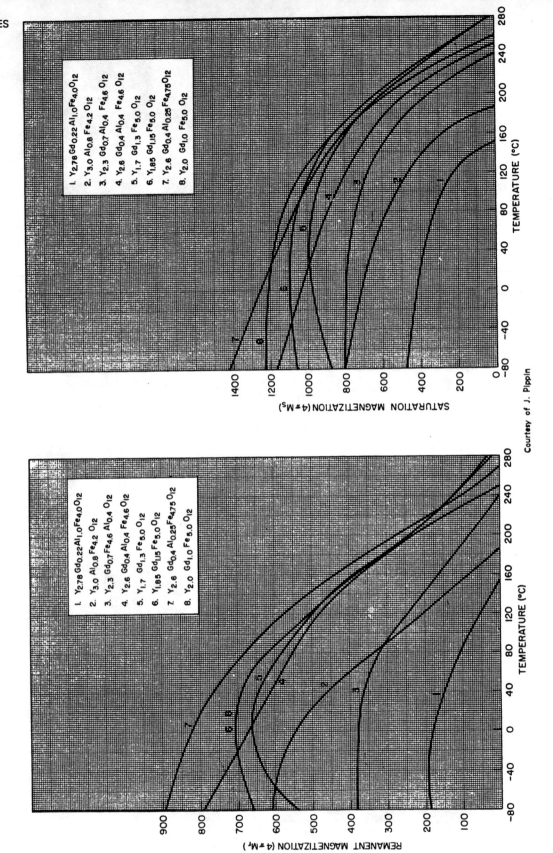

Courtesy of J. Pippin

DEMAGNETIZING FACTORS

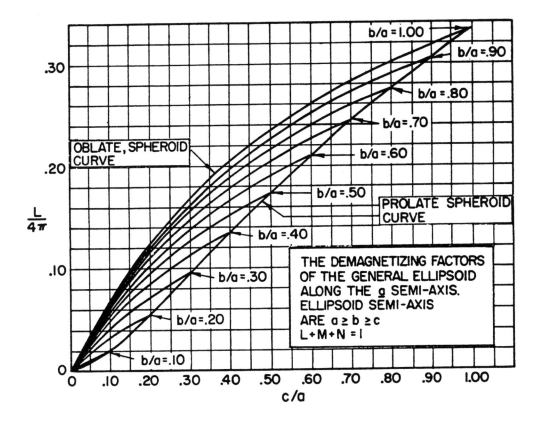

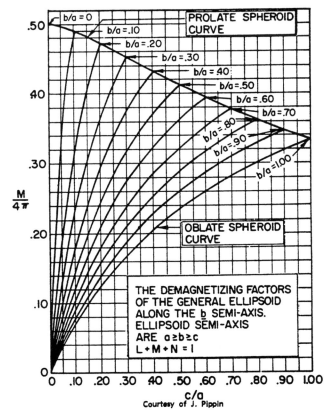

Courtesy of J. Pippin

DEMAGNETIZING FACTORS

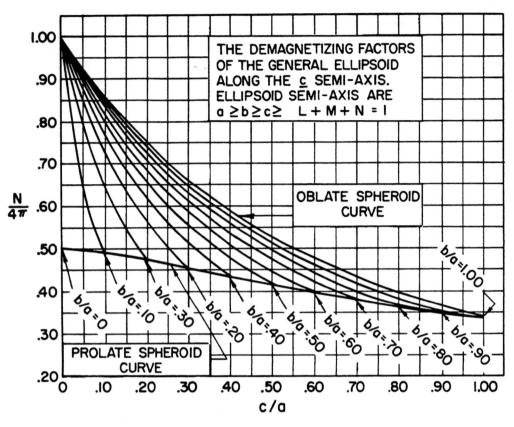

THE DEMAGNETIZING FACTORS OF THE GENERAL ELLIPSOID ALONG THE $\underline{c}$ SEMI-AXIS. ELLIPSOID SEMI-AXIS ARE $a \geq b \geq c \geq$ $L + M + N = 1$

OBLATE SPHEROID CURVE

PROLATE SPHEROID CURVE

$\frac{N}{4\pi}$

c/a

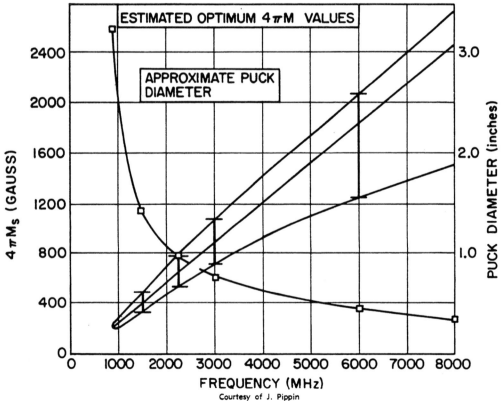

ESTIMATED OPTIMUM $4\pi M$ VALUES

APPROXIMATE PUCK DIAMETER

$4\pi M_s$ (GAUSS)

PUCK DIAMETER (inches)

FREQUENCY (MHz)

Courtesy of J. Pippin

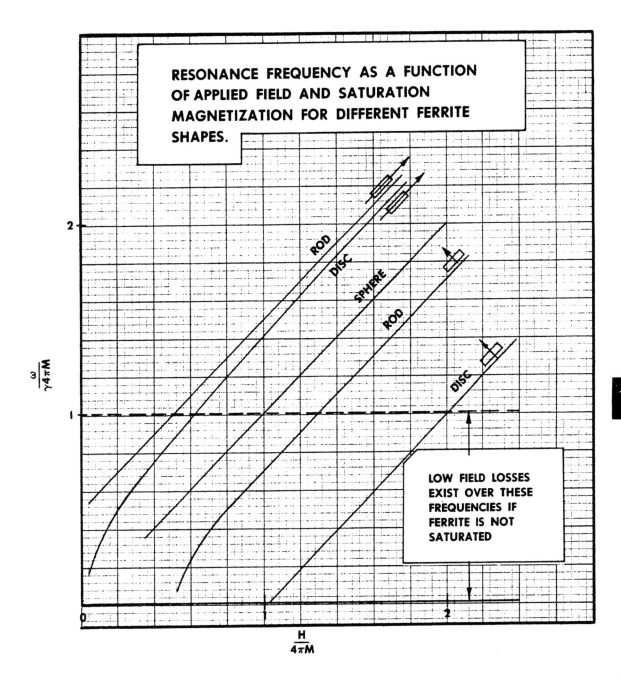

RESONANCE FREQUENCY AS A FUNCTION OF APPLIED FIELD AND SATURATION MAGNETIZATION FOR DIFFERENT FERRITE SHAPES.

LOW FIELD LOSSES EXIST OVER THESE FREQUENCIES IF FERRITE IS NOT SATURATED

J. L. Melchor and P. H. Vartanian, Reprinted from IRE Trans. on Microwave Theory and Techniques, published by the Professional Group on Microwave Theory and Techniques, Jan. 1959.

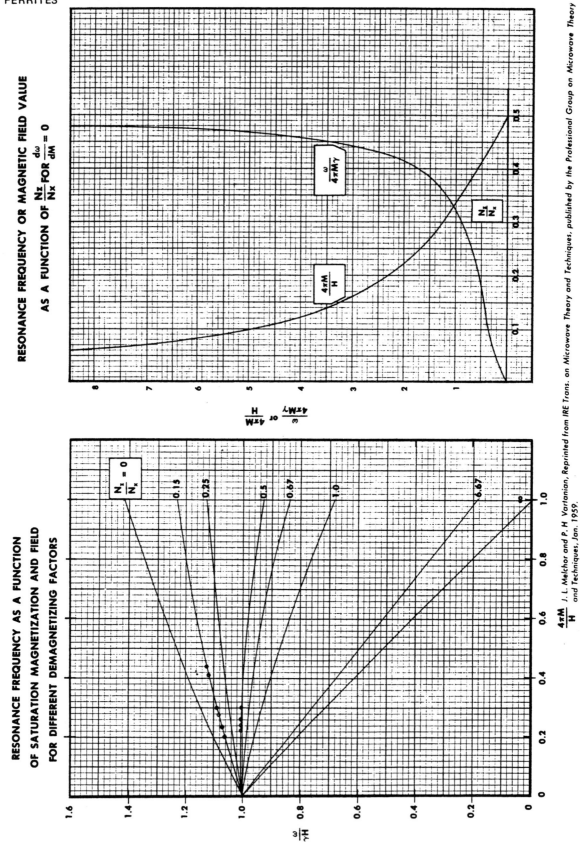

RESONANCE FREQUENCY OR MAGNETIC FIELD VALUE

AS A FUNCTION OF $\frac{N_z}{N_x}$ FOR $\frac{d\omega}{dM} = 0$

RESONANCE FREQUENCY AS A FUNCTION

OF SATURATION MAGNETIZATION AND FIELD

FOR DIFFERENT DEMAGNETIZING FACTORS

J. L. Melchor and P. H. Vartanian, Reprinted from IRE Trans. on Microwave Theory and Techniques, published by the Professional Group on Microwave Theory and Techniques, Jan. 1959.

TABLE OF THERMAL CONDUCTIVITY AND DIELECTRIC PROPERTIES OF MATERIALS COMMONLY USED IN FERRITE DEVICE DESIGN

MATERIAL	THERMAL CONDUCTIVITY (CAL/CM SEC °C)	RELATIVE DIELECTRIC CONSTANT	DIELECTRIC LOSS TANGENT
Aluminum	0.48	—	—
Ferrite (General)	0.015	12-16	—
Air	0.00010	1.00	—
Teflon	0.00059	2.02	0.004
Polystyrene	0.00024–0.00033	2.4–2.65	0.0001–0.0003
Mylar	0.000363	2.8	0.0003 (1 GHz)
Quartz	0.0033	4.4	0.0009
Boron Nitride	0.041–0.083	4.4	0.00075
Fused Silica	0.0003	3.8	0.0001–0.0004
Trans-Tech DS-6 (Fosterite)	0.009	6.5	0.0002
Beryllium Oxide	0.620	6.5	0.0005
Trans-Tech DA-9 (Alumina)	0.045	9.5	0.0002
Brush Beryllium Thermalox K (B_eO + additive)	0.310	9-25 (variable)	0.0003–0.0008
Trans-Tech D-13 (Magnesium Titanate)	0.010	13.0	0.0002
Trans-Tech D-16 (Magnesium Titanate)	0.010	16.0	0.0002

1 Cal/cm C = 2902 BTU in/hr ft^2 °F

113

Courtesy of J. Pippin

Calculated Loss of 50 Ohm Gold Microstrip Lines
Typically Used in Ferrite and Alumina Substrated Circuits

"The factor K multiplies bulk conductivity of gold. Many 'practical' microstrip circuits have relative conductivity values somewhere in the range between K=.37 and K=.75. Measured values for YIG lines agree well with D - 16 if the YIG is magnetized or for frequencies greater than approximately 5 GHz."

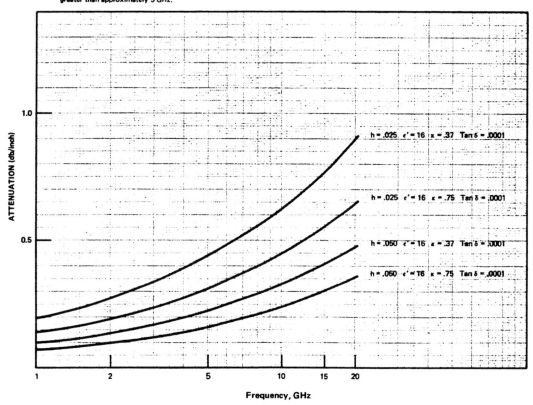

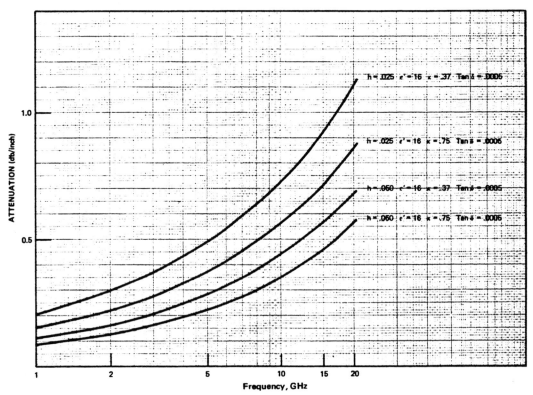

Courtesy of D.R.Taft, Sperry Microwave Electronics, Sperry Rand Corporation, Clearwater, Florida.

Calculated Loss of 50 Ohm Gold Microstrip Lines
Typically Used in Ferrite and Alumina Substrated Circuits

"The factor K multiplies bulk conductivity of gold. Many 'practical' microstrip circuits have relative conductivity values somewhere in the range between K=.37 and K=.75. Measured values for YIG lines agree well witg D - 16 if the YIG is magnetized or for frequencies greater than approximately 5 GHz."

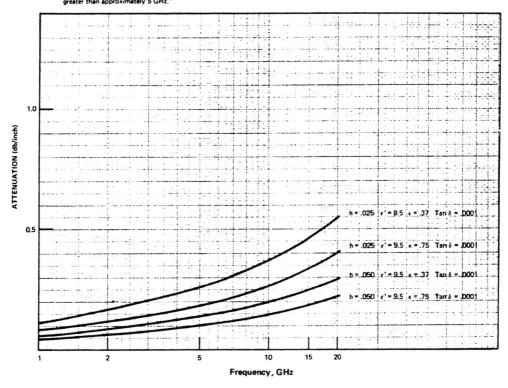

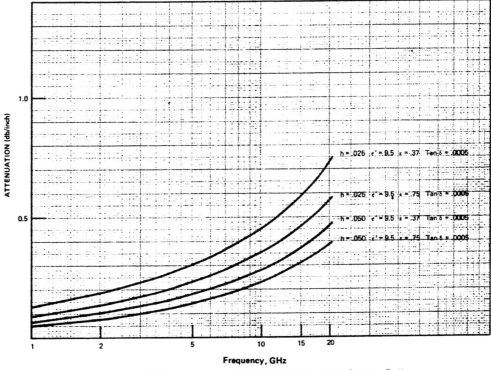

Courtesy of D.R.Taft, Sperry Microwave Electronics, Sperry Rand Corporation, Clearwater, Florida.

DIELECTRIC TRANSFORMER DESIGN

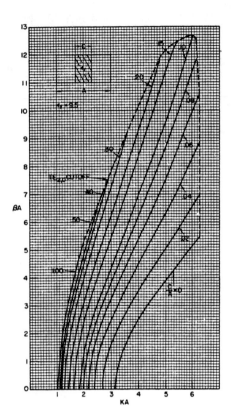

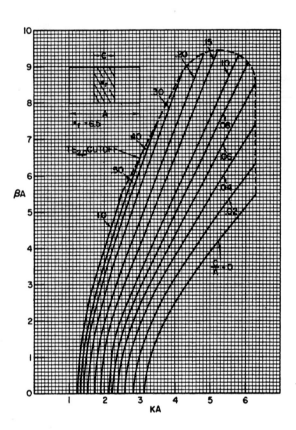

Courtesy of J. Pippin

DIELECTRIC TRANSFORMER DESIGN

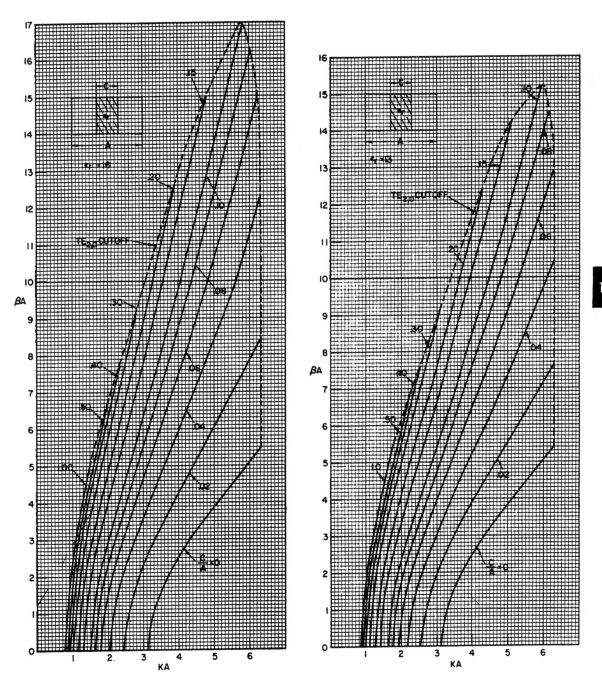

Courtesy of J. Pippin

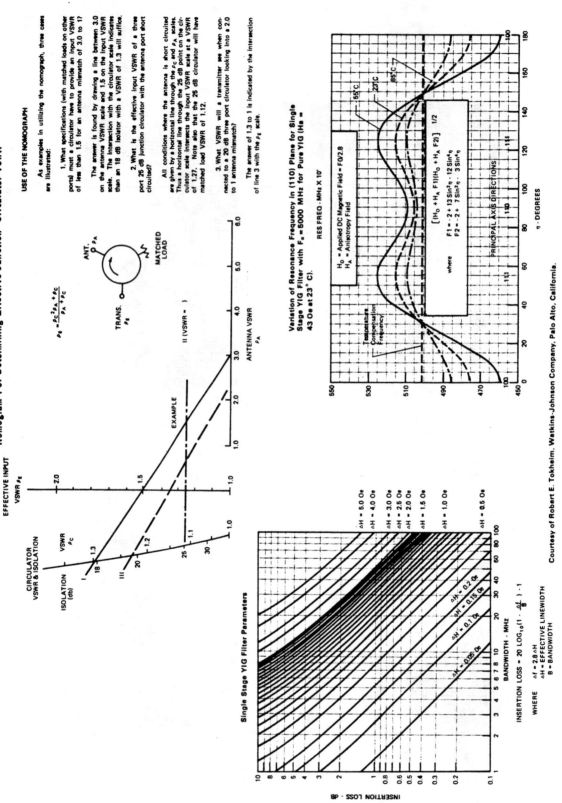

Nomogram For Determining Effective Junction Circulator VSWR

USE OF THE NOMOGRAPH

As examples in utilizing the nomograph, three cases are illustrated:

1. What specifications (with matched loads on other portal must a circulator have to provide an input VSWR of less than 1.5 for an antenna mismatch of 3.0 to 17

The answer is found by drawing a line between 3.0 on the antenna VSWR scale and 1.5 on the input VSWR scale. The intersection with the circulator scale indicates than an 18 dB isolator with a VSWR of 1.3 will suffice.

2. What is the effective input VSWR of a three port 26 dB junction circulator with the antenna port short circuited?

All conditions where the antenna is short circuited are given by a horizontal line through the ρ_C and ρ_A scales. Thus a horizontal line through the 26 dB point on the circulator scale intersects the input VSWR scale at a VSWR of 1.27. Note also that the 26 dB circulator will have matched load VSWR of 1.12.

3. What VSWR will a transmitter see when connected to a 20 dB three port circulator looking into a 2.0 to 1 antenna mismatch?

The answer of 1.3 to 1 is indicated by the intersection of line 3 with the ρ_E scale.

$$\rho_E = \frac{\rho_C^2 \rho_A + \rho_C}{\rho_A + \rho_C}$$

ANT. ρ_A

TRANS. ρ_E

MATCHED LOAD

EFFECTIVE INPUT VSWR ρ_E

CIRCULATOR VSWR & ISOLATION

ISOLATION (db)

VSWR ρ_C

ANTENNA VSWR ρ_A

II (VSWR =)

EXAMPLE

Variation of Resonance Frequency in (110) Plane for Single Stage YIG Filter with F_o = 5000 MHz for Pure YIG (Ha = 43 Oe at 23°C).

RES FREQ - MHz X 10³

H_O — Applied DC Magnetic Field = F0/2.8
H_A — Anisotropy Field

$$[(H_O + H_A \, F)](H_O + H_A \, F2)]^{1/2}$$

where
F1 = -2 + 13 Sin²η - 12 Sin⁴η
F2 = -2 + 7 Sin²η - 3 Sin⁴η

Temperature Compensation Frequency

55°C
23°C
85°C

PRINCIPAL AXIS DIRECTIONS

η - DEGREES

Single Stage YIG Filter Parameters

BANDWIDTH - MHz

INSERTION LOSS = 20 LOG₁₀(1 - $\frac{\Delta f}{B}$) - 1

WHERE Δf = 2.8 ΔH
ΔH = EFFECTIVE LINEWIDTH
B = BANDWIDTH

INSERTION LOSS - dB

ΔH = 5.0 Oe
ΔH = 4.0 Oe
ΔH = 3.0 Oe
ΔH = 2.5 Oe
ΔH = 2.0 Oe
ΔH = 1.5 Oe
ΔH = 1.0 Oe
ΔH = 0.5 Oe
ΔH = 0.2 Oe
ΔH = 0.15 Oe
ΔH = 0.1 Oe
ΔH = 0.05 Oe

Courtesy of Robert E. Tokheim, Watkins-Johnson Company, Palo Alto, California.

IMPORTANT RELATIONSHIPS

1. Kittel's equation for resonance

$$f = \gamma \left\{ [H_A - (N_Z - N_X)M] [H_A - (N_Z - N_Y)M] \right\}^{\frac{1}{2}}$$

where f = resonant frequency in megacycles
  γ = gyromagnetic ratio = 2.8 oersteds per megacycle
  H_A = applied steady stage magnetic field (in Z direction)
  M = magnetization
  N_X, N_Y, N_Z = demagnetizing factors. $(N_X + N_Y + N_Z = 4\pi)$

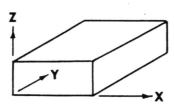

2. Back-to-front ratio, R, of resonant isolator

$$R = \left(\frac{4f}{\gamma \Delta H} \right)^2 \qquad \text{where } \Delta H = \text{line width}$$
$$\qquad\qquad\qquad f \text{ and } \gamma \text{ as above}$$

3. Figure of merit, F, for off-resonant devices

$$F = \frac{2f}{\gamma \Delta H} \text{ radians per nepers}$$

Courtesy of David Andrews.

NOISE FIGURE MEASUREMENTS

The sensitivity of amplifiers or receiving systems is usually expressed in terms of one or more of the three noise parameters: noise figure; effective input noise temperature; operating noise temperature. Figure 1 is a noise model of a simple receiver illustrating these parameters.

Ta	= Antenna Temperature, Kelvins	G	= Power Gain (Ratio)
Top	= Operating Noise Temperature, Kelvins	No	= Output Noise Power, Watts
Te	= Effective Input Noise Temp., Kelvins	K	= Boltzmann's Constant
F	= Noise Figure (Ratio)		= 1.38×10^{-23} Joules/Kelvin
To	= 290 Kelvins	B	= Noise Bandwidth

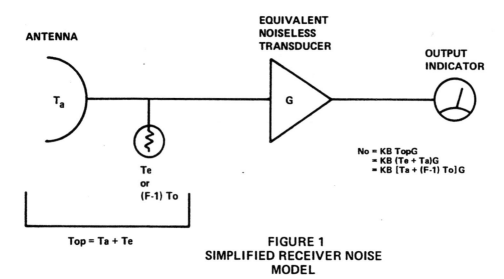

$$Top = Ta + Te$$

FIGURE 1
SIMPLIFIED RECEIVER NOISE MODEL

The three noise parameters are usually measured by means of the "Y-factor" method. Figure 2 illustrates the procedure for noise figure and effective input noise temperature.

PROCEDURE

1. Noise generator at T_1 ($T_1 < T_2$); adjust amp-det gain for convenient meter reference.
2. Switch noise generator to T_2; adjust attenuator for the same meter reference.
3. Note change in attenuator setting, Δ A(dB)

4. Calculate, $Y = 10^{\dfrac{-\Delta A}{10}} = \dfrac{Log^{-1} \Delta A}{10}$
5. Calculate noise figure or effective input noise temperature from equations below or from the curves on the next pages.

$$F = \frac{\dfrac{T_2}{To} - Y\dfrac{T_1}{To}}{Y-1} + 1 \qquad Te = \frac{T_2 - Y\,T_1}{Y-1}$$

If $T_1 = To$

$$F = \frac{\left(\dfrac{T_2}{To} - 1\right)}{Y-1}$$

$$ENR = EXCESS\ NOISE\ RATIO = \left(\frac{T_2}{To} - 1\right)$$

$$F_{dB} = ENR_{dB} - 10\ Log\ (Y-1)$$

FIGURE 2
MEASUREMENT OF F AND Te

Courtesy of William Pastore

Figures 3 and 4 plot F and T_e vs. Y-factor for two types of noise sources. Figure 3 is used with thermal noise generators using liquid nitrogen and boiling water temperatures. Figure 4 plots noise figure for a range of excess noise ratios typical of gas-discharge noise generators.

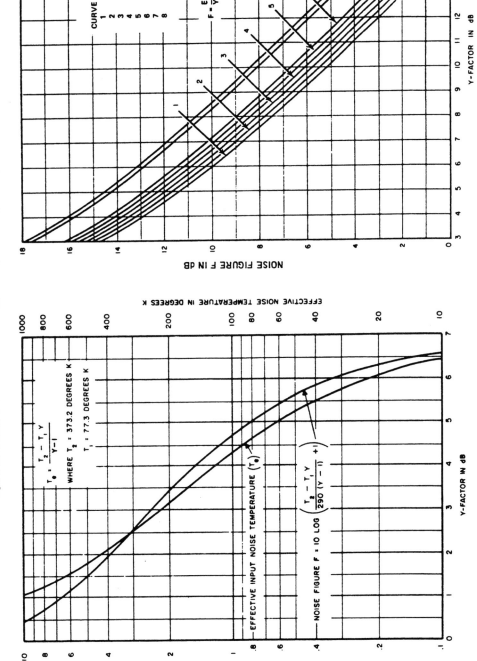

FIGURE 4. MEASURED NOISE FIGURE VS Y-FACTOR FOR T_1= 290K

FIGURE 3. NOISE FIGURE AND EFFECTIVE INPUT NOISE TEMPERATURE VS Y-FACTOR FOR TERMINATION TEMPERATURES OF 77.3°K AND 373.2°K

Courtesy of William Pastore

123

When coaxial Hot/Cold terminations are used as the noise source, losses in the coaxial line can cause the output noise temperature to differ from the thermal temperatures. Figure 5 is a plot of these differences for a typical Hot/Cold noise source (AIL 07009).

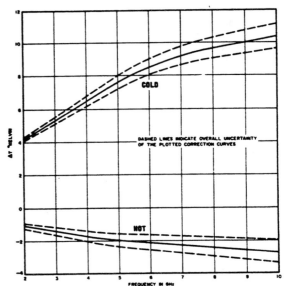

FIGURE 5. CORRECTIONS IN HOT AND COLD TERMINATION TEMPERATURES DUE TO INTERNAL RF LOSSES FOR THE AIL 07009 HOT/COLD NOISE GENERATOR.

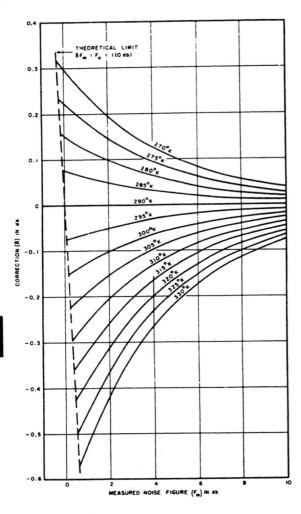

FIGURE 6. CORRECTION FOR T_1 To WHEN T_2=10,100K

If a room temperature termination is used as the cold source, and the simplified relation for noise figure is employed, the result is in error. Figure 6 is a plot of the correction to be applied for a typical argon gas-discharge noise generator.

The excess noise ratio specified for gas-discharge noise generators is usually that of the tube alone. The effect of the microwave coupling structure on the noise output is related to the hot and cold (on off respectively) insertion loss. The reduction in the excess noise ratio of the tube due to these insertion losses is shown in Figure 7.

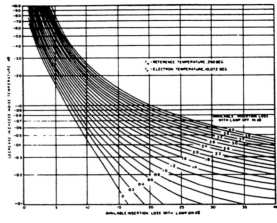

FIGURE 7. DECREASE IN RELATIVE EXCESS NOISE TEMPERATURE AS A FUNCTION OF NOISE SOURCE INSERTION LOSS

Present day solid state noise sources have excess noise ratios as high as 35 dB in some cases. This has made the measurement of operating noise temperature as convenient and accurate as a noise figure measurement. The Y-factor method is still used; however, the noise is injected by means of a directional coupler located just after the antenna. Figure 8 shows the setup, and Figure 9 is a plot of the measurement equation.

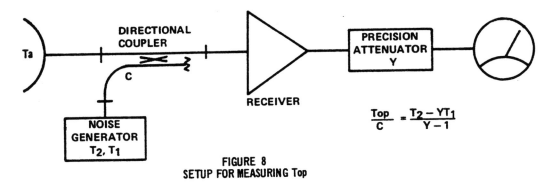

$$\frac{T_{op}}{C} = \frac{T_2 - YT_1}{Y - 1}$$

FIGURE 8
SETUP FOR MEASURING Top

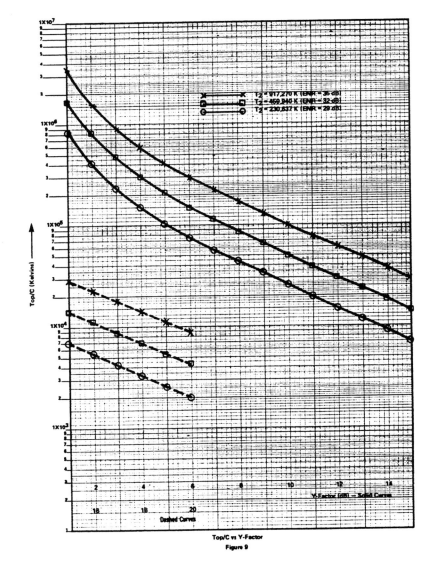

Top/C vs Y-Factor
Figure 9

Courtesy of William Pastore

NOISE FIGURE MEASUREMENTS

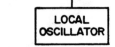

| TERMINATION | NOISE GENERATOR | AMPLIFIER UNDER TEST | ⊗ MIXER | PRECISION IF ATTENUATOR | POSTAMPLIFIER DETECTOR AND METER |

LOCAL OSCILLATOR

IF ATTENUATOR METHOD ("Y" - FACTOR METHOD)

This is the most accurate method known. When used in this system, the Precision Attenuator will yield Y-factor measurements that are accurate to within 0.1 db. Nonlinearities in the components following the attenuator are of no consequence, since the attenuator is used to adjust signal level to the same meter reading when the noise generator is in the on and off conditions. Noise figures can be measured accurately using the equation:

$$F_{db} = 10 \log (T_d/T_o - 1) - 10 \log (Y - 1)$$

where

T = temperature of noise generator (known)
T_o = 290°K
Y = IF attenuation, or ratio of signal power when noise generator is on to signal power when noise generator is off.

A widely used method of obtaining Y-factor involves the use of a precision attenuator in the IF system. These attenuators are readily available commercially, and they can be obtained with extremely high accuracy. When using an IF attenuator, all effects of circuits after the attenuator are eliminated since the output is always adjusted to the same level. The attenuator is adjusted to maintain the receiver output constant for any condition of receiver input, and the change in the receiver output is indicated by the change in the attenuator setting.

The definition of noise figure refers the measurement to room temperature (290°K). The measurement seldom, if ever, is made at that temperature. Specifically, in a well-designed diode noise gener-

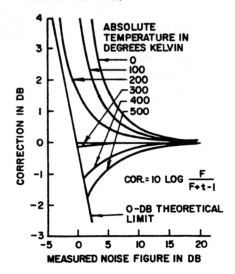

ator, the terminating resistor is close to the tube and, therefore, becomes hotter than room temperature. For high noise figures, the correction is small and is usually neglected. With low noise figures, on the other hand, the correction is significant. A plot of the temperature correction is shown.

Courtesy of C. L. Cuccia

HOT — COLD NOISE FIGURE MEASUREMENTS

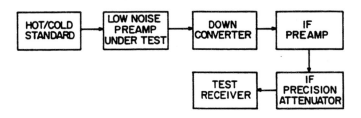

| HOT/COLD STANDARD | LOW NOISE PREAMP UNDER TEST | DOWN CONVERTER | IF PREAMP |

| TEST RECEIVER | IF PRECISION ATTENUATOR |

The standard is switched between the precision hot and cold bodies. The IF attenuator is adjusted to obtain the same detector reading for the hot and cold bodies. This will yield Y-factor measurements that are accurate to within 0.1 db. Non-linearities in the detector are of no consequence, since the precision attenuator keeps the detector at the same operating point. This method is particularly useful and accurate for noise figures under 5 db and is the only method with proper accuracy for noise figures under 3 db.

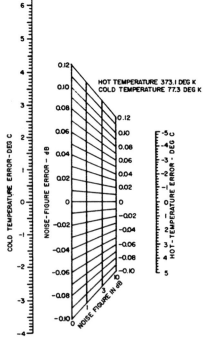

NOISE FIGURE ERROR AS A FUNCTION OF HOT AND COLD TEMPERATURE ERRORS

Courtesy of A. I. L.

NOISE FIGURE MEASUREMENT

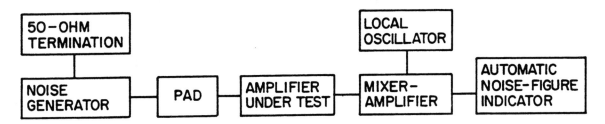

This method eliminates inaccuracies caused by variations in the gain of the receiver under test, and since noise figure is instantaneously displayed by the meter, it is much quicker than other methods. If a means of controlling amplifier gain and a means of turning the noise source on and off manually are provided, the system can be calibrated quickly and accurately by inserting the IF attenuator after the receiver and using the IF attenuator method. (Y-factor method).

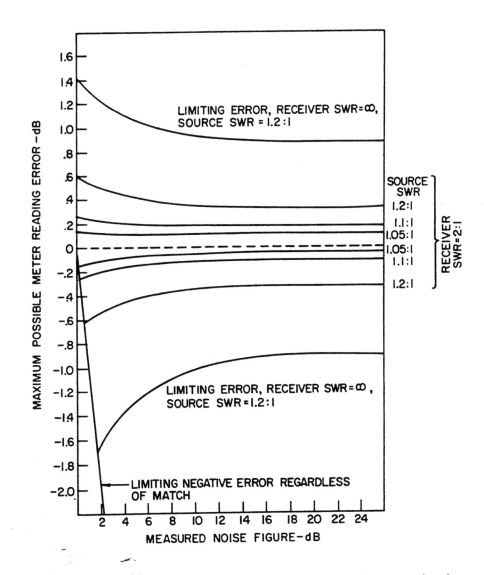

Typical error effects for several possible conditions of mismatch between noise source and receiver.

Courtesy Hewlett Packard

VSWR Effects in Noise - Figure Measurements

Some of the parameters that affect the accuracy of a noise-figure measurement are:

1. The fired and unfired VSWR of the noise generator.

2. The VSWR of the receiver or device under test.

3. The noise figure of the receiver or device under test.

Since each of these present an interacting effect on each other, computing the error for a given set of conditions involves a lengthy mathematical procedure.

The attached nomograph presents a rapid means of computing the limits within which the error will lie. It must be stated at the outset that we have made one assumption that could make the error even greater than the limits that will be indicated on the nomograph. This assumption is that the receiver gain and noise figure will remain constant for both hot and cold VSWR conditions of the noise generator. This is not always true particularly in the case of low noise and/or negative resistance input amplifiers. For this reason the nomograph has been limited to receivers whose noise figure is greater than 5 db.

The use of the nomograph is best shown by a typical example. Assume the following conditions:

Noise Generator VSWR–Fired	2.0
Noise Generator VSWR–Unfired	3.0
Receiver VSWR	2.0
Receiver Noise Figure	7 db

Referring to Figure 1, "Interaction K Factors vs VSWR of Receiver and Noise Generator," plot the intersecting points of the Receiver VSWR (2.0) and both the Fired and Unfired Noise Generator VSWR (2.0 and 3.0). From each point, read two K Factors–one from the solid family of lines, the other from the dashed family of lines (see locations marked X on the graph). For this example, the Fired K Factors are 1.125 and 0.725 and the Unfired K Factors are 1.08 and 0.56.

Next refer to Figure 2 and plot the Unfired K points. Extend these points, following the general family of curves, to the point where they intersect with the 7-db noise-figure line. These points are then projected horizontally back to the Unfired K line. Next plot the Fired K points on the scale on the left. Connect these points to the previously projected points on the Unfired K line in such a manner that the lines do not corss. The points where these lines intersect the middle line are the limits of the error of the noise-figure measurement. In this case, the error can be from +1.45 to −0.90 db. These are the limits of the error created by the impedance characteristics only. Other errors such as noise generator tolerance, cable losses, indicator tolerances, etc., are in addition to this error. Also, as previously indicated, should the receiver gain and/or noise figure vary with change in VSWR conditions of the Noise Generator, the error can be even greater than the nomograph indicates.

128

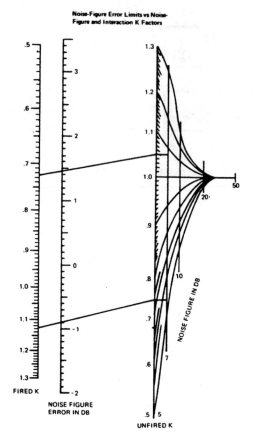

Noise-Figure Error Limits vs Noise-Figure and Interaction K Factors

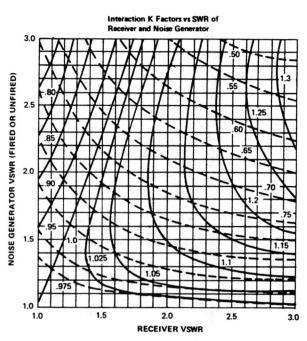

Interaction K Factors vs SWR of Receiver and Noise Generator

Courtesy of AIL, A Division of Cutler - Hammer, Inc. Deer Park, New York.

When measuring the noise figure of an amplifier, it is often necessary to insert additional gain between the device under test and the attenuator. The measured noise figure is the overall noise figure of the cascade. The nomograph shown below provides a means of determing the first stage noise figure when its gain and the noise figure of the second stage are known.

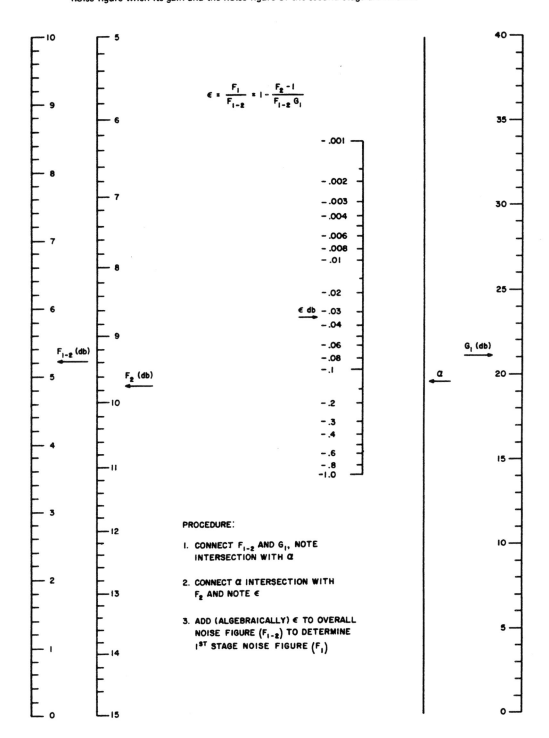

$$\epsilon = \frac{F_1}{F_{1-2}} = 1 - \frac{F_2 - 1}{F_{1-2}\,G_1}$$

PROCEDURE:

1. CONNECT F_{1-2} AND G_1, NOTE INTERSECTION WITH α

2. CONNECT α INTERSECTION WITH F_2 AND NOTE ϵ

3. ADD (ALGEBRAICALLY) ϵ TO OVERALL NOISE FIGURE (F_{1-2}) TO DETERMINE 1ST STAGE NOISE FIGURE (F_1)

NOMOGRAPH, CORRECTION FOR SECOND STAGE NOISE FIGURE

Courtesy of William Pastore

GRAPH OF NOISE CORRECTION vs. HOT INSERTION LOSS OF GAS
DISCHARGE NOISE GENERATORS

Let L db = 10 log L', where L' is the actual power ratio represented by L db. The value of the correction, Δ is L/(L-1), expressed in db and may be calculated from L db - (L-1) db = Δdb.

To provide rapid determination of the correction to the excess noise ratios from tube only-to-tube-and-mount, the graph provides the values of the correction, L/(L-1) db, as a function of hot insertion loss, L db.

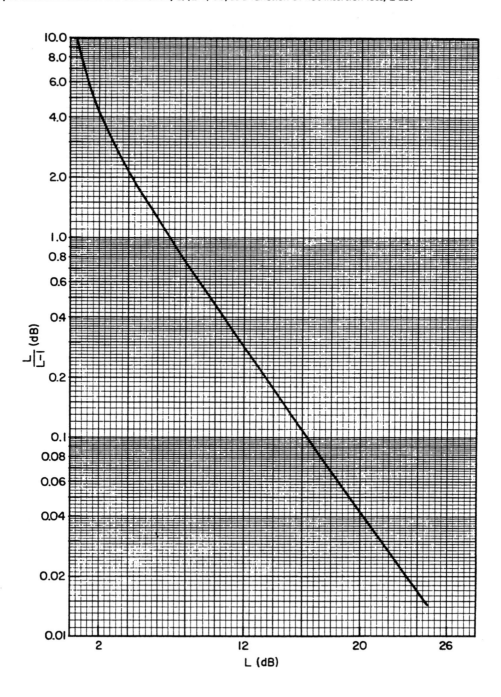

Courtesy K. W. Olson and B. A. Wheeler, Signalite Inc., Special Products Division, Neptune, New Jersey, a General Instrument Company.

NOISE TEMPERATURE EQUATIONS FOR COMMONLY USED CIRCUITS

CASE I: SIGNIFICANT LOSS – CIRCUITS BETWEEN ANTENNA & AMPLIFIER MIXER

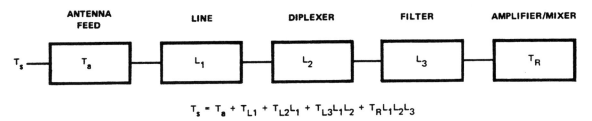

$$T_s = T_a + T_{L1} + T_{L2}L_1 + T_{L3}L_1L_2 + T_RL_1L_2L_3$$

CASE II: SIGNIFICANT LOSS – CIRCUITS BETWEEN AMPLIFIERS AND MIXER/RECEIVER

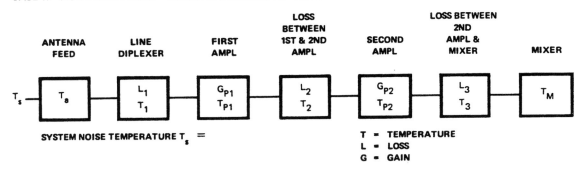

SYSTEM NOISE TEMPERATURE T_s =

T = TEMPERATURE
L = LOSS
G = GAIN

$$T_a + T_1(L_1 - 1) + T_{P1}L_1 + T_2 \frac{L_1}{G_{P1}} (L_2 - 1) + T_{P2} \frac{L_1L_2}{G_{P1}} + T_3 (L_3 - 1) \frac{L_1L_2}{G_{P_1}G_{P_2}} + T_M \frac{L_1L_2L_3}{G_{P_1}G_{P_2}}$$

CASE III: SIGNIFICANT ANTENNA – COMPARATOR VSWR SEEN BY FIRST AMPLIFIER

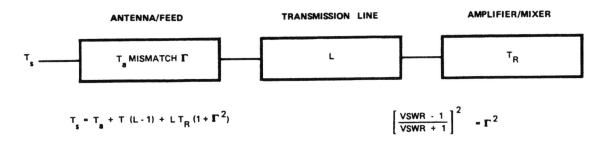

$$T_s = T_a + T (L - 1) + L T_R (1 + \Gamma^2)$$

$$\left[\frac{VSWR - 1}{VSWR + 1} \right]^2 = \Gamma^2$$

CASE IV: AMPLIFIER MIXER CHAIN

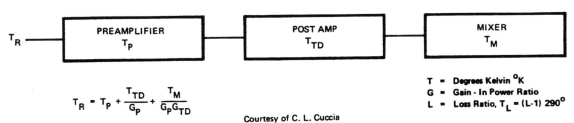

$$T_R = T_P + \frac{T_{TD}}{G_P} + \frac{T_M}{G_PG_{TD}}$$

T = Degrees Kelvin °K
G = Gain - In Power Ratio
L = Loss Ratio, $T_L = (L-1) 290°$

Courtesy of C. L. Cuccia

NOISE FIGURE OF MICROWAVE AMPLIFIERS

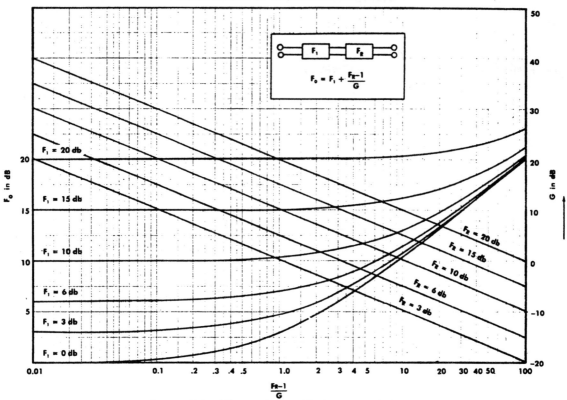

Courtesy of Polarad Electronics Corp., modified by W. W. Mumford, B. T. L.

OVERALL NOISE FIGURE NOMOGRAPH
WHEN 10 LOG F_M = 1.5 dB

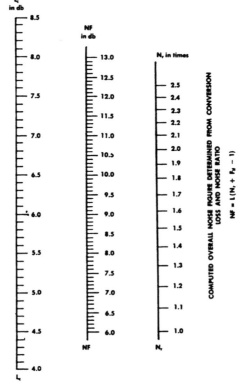

Courtesy of Sylvania Electric Products, Inc.

NOISE GENERATOR CALIBRATION SERVICES
AVAILABLE AT NATIONAL BUREAU OF STANDARDS

The accuracy of any noise figure or noise temperature measurement is limited by the accuracy of the noise source. Calibration equipment and services are available for noise generators over a large portion of the R.F. spectrum. The table below lists the services available, as of this date, at the Boulder laboratories of the National Bureau of Standards .

1. COAXIAL NOISE GENERATORS

Frequency	Noise Temperature Range	Limit of Uncertainty
MHz	K	K
3	75 to 30,000	±1%

2. WAVEGUIDE NOISE GENERATORS

Frequency Range GHz	Waveguide Size	Flange	ENR Range dB	Limits of Uncertainty dB
12.4 - 18.0	WR-62 (RG-91/U)	UG-419/U	4 - 30	±0.06 - 0.08
10.0 - 15.0	WR-75	Cover	4 - 30	±0.07 - 0.09
8.2 - 12.4	WR-90 (RG-52/U)	UG-39/U	4 - 30	±0.06 - 0.08
8.2 - 10.0	WR-112 (RG-51/U)	UG-51/U	4 - 30	±0.07 - 0.09
3.3 - 3.95	WR-229	Cover	1.5 - 30	±0.13 - 0.15
2.6 - 3.95*	WR-284 (RG-48/U)	UG-53/U	1.5 - 30	±0.12 - 0.14

*Also coaxial noise generators fitted with 14 mm. precision connectors.

133

LOW LEVEL DETECTION CHARACTERISTICS

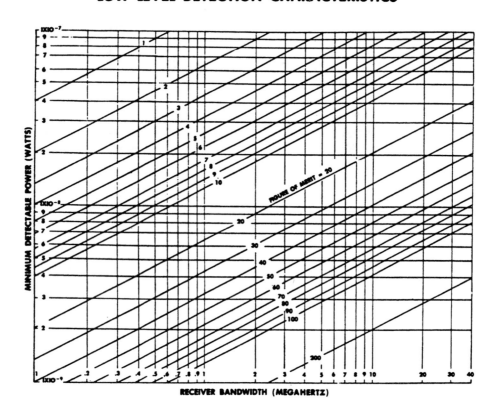

LINEAR DETECTOR, INCREASE OF AVERAGE OUTPUT DUE TO NOISE

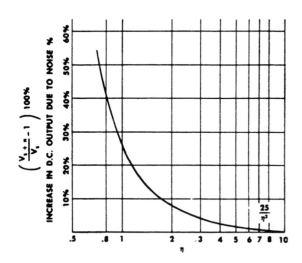

$$\% \text{ INCREASE} = \left\{ \frac{1}{2\sqrt{\pi\eta}} \cdot e^{\frac{-\eta^2}{2}} \left[(1+\eta^2) I_o \left(\frac{\eta^2}{2}\right) + \eta^2 I_1 \left(\frac{\eta^2}{2}\right) \right] - 1 \right\} 100\%$$

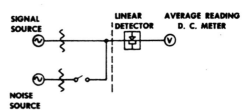

η = RATIO OF RMS VOLTAGES OF SIGNAL AND NOISE
    AT INPUT OF LINEAR DETECTOR
V_{s+n} = AVERAGE READING DUE TO SIGNAL PLUS NOISE
V_s = AVERAGE READING DUE TO SIGNAL ALONE
$I_n(x)$ IS A MODIFIED BESSEL FUNCTION OF nth ORDER

Top: *Courtesy of Sylvania Electric Products, Inc.*   Bottom: *Courtesy of Weinschel Engineering Co., Inc.*

FLUCTUATION OF LINEAR DETECTOR OUTPUT

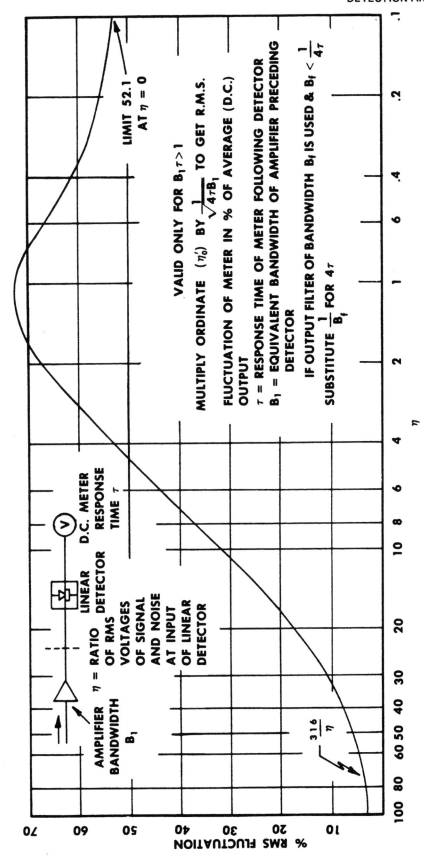

Courtesy of Weinschel Engineering Co., Inc.

135

Nomogram for Video Detector Tangential Sensitivity

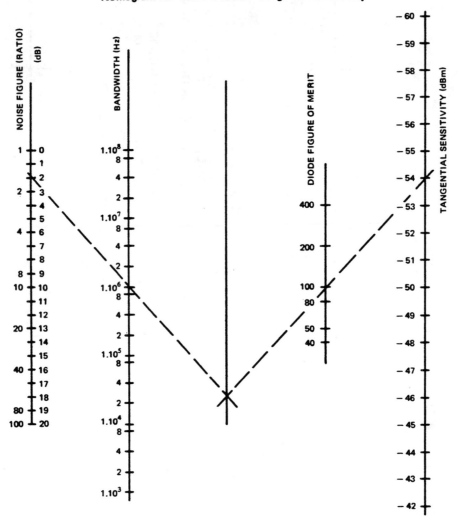

EXAMPLE: FIND TANG. SENSITIVITY FOR
VIDEO DETECTOR HAVING FIG. OF MERIT = 100
OPERATING INTO AN AMPLIFIER HAVING NF = 2 dB
AND BW = 1 mHz.

The accepted criterion for comparison of detector performance is, and has been, Tangential-Signal Sensitivity. Due to the subjective nature of this test, correlation of results has been unsatisfactory.

A less subjective test of detector performance is made by measuring the Figure of Merit. The Figure of Merit is related to Tangential-Signal Sensitivity through an equation involving the video-amplifier Noise Figure and Bandwidth.

To calculate the Figure of Merit for a detector, the open-circuit voltage sensitivity (γ = millivolts per milliwatt) and the video resistance (R_v) are measured. The readings are taken at a low enough RF-power level to insure operation in the square-law region. The Figure of Merit (M) then equals $\gamma/R_v^{\frac{1}{2}}$. Knowing the amplifier Noise Figure (F) and Bandwidth (B), the Tangential Signal Sensitivity can be calculated using the equation.

$$TSS (dBm) = -35 + 5 \log (BxF) - 10 \log M$$

It is important in this expression to use (1) the Noise Figure of the amplifier when driven with the diode video resistance, and (2) the amplifier Noise Bandwidth. For example, the Noise Bandwidth of an amplifier with a 6 dB/octave rolloff equals 1.51 times the 3-dB Bandwidth.

The attached nomogram provides convenient method of computing TSS. The use of the nomogram is self-explanatory. It allows solving for any one of the four unknowns, given the other three.

Courtesy of R. Ferri, Microphase Corporation, Greenwich, Connecticut.

RECEIVER SENSITIVITY NOMOGRAPH

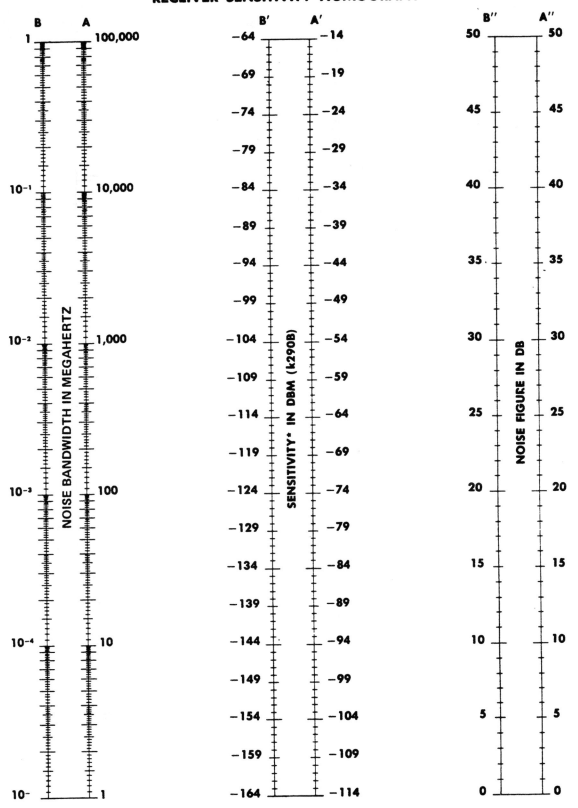

*Assuming an output signal-to-noise ratio of unity when the generator temperature is 290° Kelvin.

Courtesy of J. C. Stevenson, Huggins Laboratories Inc., Sunnyvale, Cal.

Receiver R.M.S. Absolute Noise Threshold

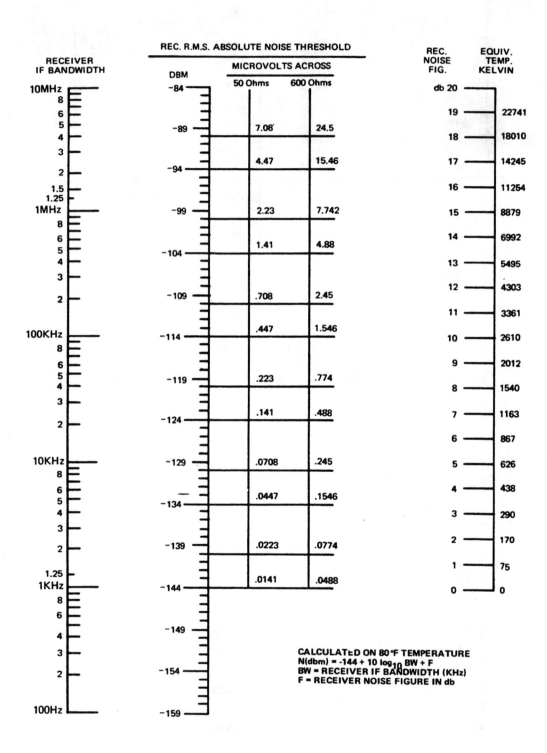

REC. R.M.S. ABSOLUTE NOISE THRESHOLD

CALCULATED ON 80°F TEMPERATURE
$N(dbm) = -144 + 10 \log_{10} BW + F$
BW = RECEIVER IF BANDWIDTH (KHz)
F = RECEIVER NOISE FIGURE IN db

Courtesy of R. L. Marks, F. Zawislan, Lt. J. McClure, USAF,
Lt. R. Fellows, Jr., USAF, Rome Air Development Center

Noise Figure vs. Noise Temperature

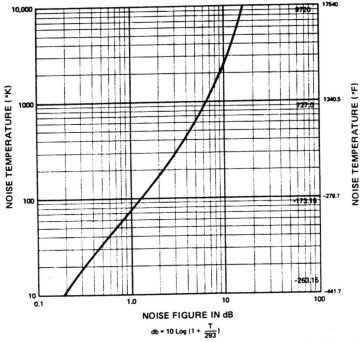

NOISE FIGURE IN dB

$$db = 10 \log \left(1 + \frac{T}{293}\right)$$

Courtesy of R.L.Marks, F.Zawislan, Lt. J. McLure, USAF, Lt. R.Fellows, Jr., USAF, Rome Air Development Center, N.Y.
Reference: "Some Aspects of FM Design for Line - of - sight Microwave and Troposcatter Systems," technical report no.
RADC - TR - 65 - 51, April 1965.

SNR Degradation Due To Multiplier Noise in Multiplier Chains

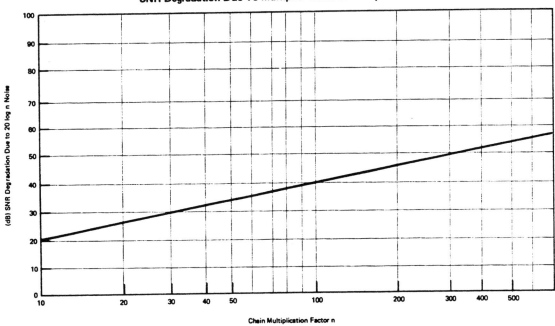

139

Noise Power – Noise Unit Conversion Chart
(1 MW – 0 Level Reference)

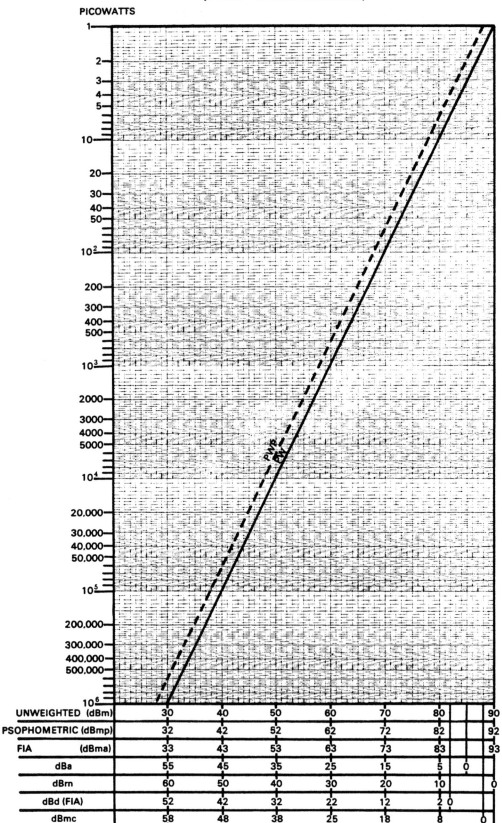

	30	40	50	60	70	80			90
UNWEIGHTED (dBm)	30	40	50	60	70	80			90
PSOPHOMETRIC (dBmp)	32	42	52	62	72	82			92
FIA (dBma)	33	43	53	63	73	83			93
dBa	55	45	35	25	15	5	0		
dBrn	60	50	40	30	20	10			0
dBd (FIA)	52	42	32	22	12	2	0		
dBmc	58	48	38	25	18	8		0	

Courtesy of Leo Joergensen, Federal Electric Corporation, Paramus, N.J.

Conversion of S + N/N (dB) to S/N (dB) Addition of Noise

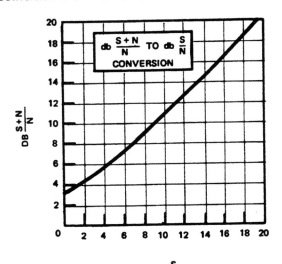

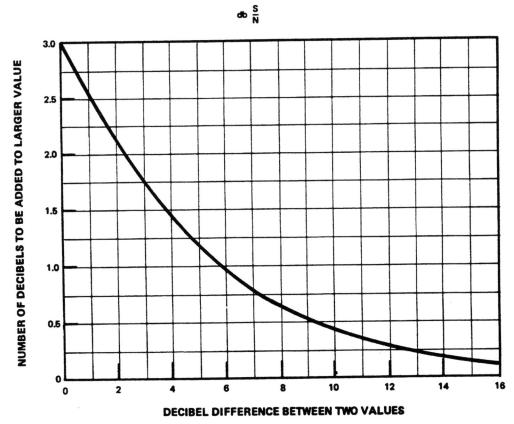

NUMBER OF DECIBELS TO BE ADDED TO LARGER VALUE

DECIBEL DIFFERENCE BETWEEN TWO VALUES

G/T Nomograph for 4 GHz

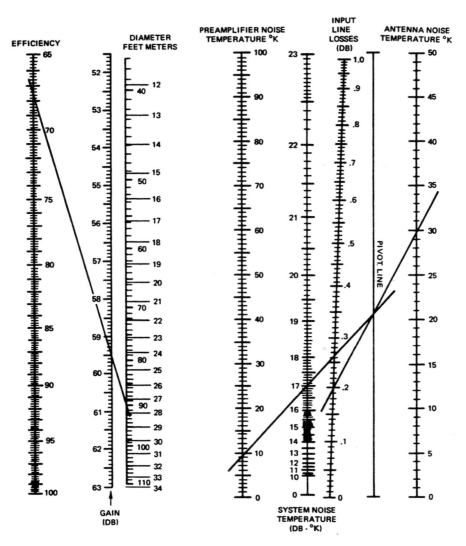

EFFICIENCY

DIAMETER FEET METERS

GAIN (DB)

PREAMPLIFIER NOISE TEMPERATURE °K

INPUT LINE LOSSES (DB)

ANTENNA NOISE TEMPERATURE °K

PIVOT LINE

SYSTEM NOISE TEMPERATURE (DB - °K)

TO USE NOMOGRAPH:

1. Lay straight edge through known antenna diameter and efficiency. This yields gain of antenna at feed input. (See line 1 of example.) To find gain at preamplifier input, subtract line losses (in db) from antenna gain.
2. Lay straight edge through antenna noise temperature and input line loss. Note where straight edge intersects vertical pivot line. (See line 2 of example.)
3. From above intersection of straight edge and pivot line, lay straight edge through preamplifier noise temperature. This yields system noise temperature in db as measured at input to preamplifier. (See line 3 of example.)
4. Subtract noise temperature (in db) as determined in step 3 from gain as determined in step 1. This gives antenna G/T at the specified elevation angle.

Example: Antenna diameter = 90 ft. } Gain=59.5 db (from line 1)
Efficiency =67%

Antenna Noise Temperature (at 10° elevation) =30° } System Noise Temperature=17.1 db
Input Line Losses =0.2 db } (from lines 2 and 3)
Preamplifier Noise Temperature =10°K

G/T=59.5 - .2 (line losses) – 17.1=42.2 db

Courtesy of Kazu - Oshima, Philco - Ford, Palo Alto, California. Reference: Cuccia, C.L., T.G.Williams, P.R.Cobb, A.E.Smoll, and J.P.Rahilly, "RF Design of Communication Satellite Earth Stations", Microwaves, May 1967, Part I.

Instantaneous Addition of Signal and Noise in a Dual Diversity Combiner

$$\frac{S_1}{N_1} + \frac{S_2}{N_2} = \frac{S_1 + K S_2}{\sqrt{(N_1)^2 + (K N_2)^2}}$$

MAXIMAL RATIO $K = \frac{S_2}{S_1}$

EQUAL GAIN [K = 1]

SWITCH [K = 0]

db DIFFERENCE IN S/N RATIO AT COMBINER INPUT

db CONTRIBUTION OF THE SMALLER S/N RATIO TO THE COMBINED OUTPUT

Courtesy of R.L.Marks, F.Zawislan, Lt. J. McLure, USAF, Lt. R.Fellows, Jr., USAF, Rome Air Development Center, N.Y.
Reference: "Some Aspects of FM Design for Line - of - sight Microwave and Troposcatter Systems," technical report no.
RADC - TR - 65 - 51, April 1965.

143

Efficiency of Nonmatched Filters
Compared with the Matched Filter

Input Signal	Filter	Loss in SNR compared with matched filter, db
Rectangular pulse	Rectangular	.9
Rectangular pulse	Gaussian	0.98
Gaussian pulse	Rectangular	0.98
Gaussian pulse	Gaussian	0 (matched)
Rectangular pulse	One stage, single-tuned circuit	0.88
Rectangular pulse	2 cascaded single-tuned stages	0.56
Rectangular pulse	5 cascaded single tuned stages	0.5

Comparison of Noise Bandwidth and 3 db Bandwidth

Type of receiver coupling circuit	No. of stages	Ratio of noise bandwidth to 3-db bandwidth
Single-tuned	1	1.57
	2	1.22
	3	1.16
	4	1.14
	5	1.12
Double-tuned $^{++}$	1	1.11
	2	1.04
Staggered triple	1	1.048
Staggered quadruple	1	1.019
Staggered quintuple	1	1.01
Gaussian	1	1 065

++ Applies to a transitionally coupled double-tuned circuit or to a stagger-tuned circuit with
two tuned circuits

From Introduction to Radar Systems by Merrill J. Skolnik. Copyright McGraw - Hill Inc., 1962.
Used with permission of McGraw - Hill Book Company.

Microwave Engineers'

Probability of Bit Error vs Signal to Noise Ratio

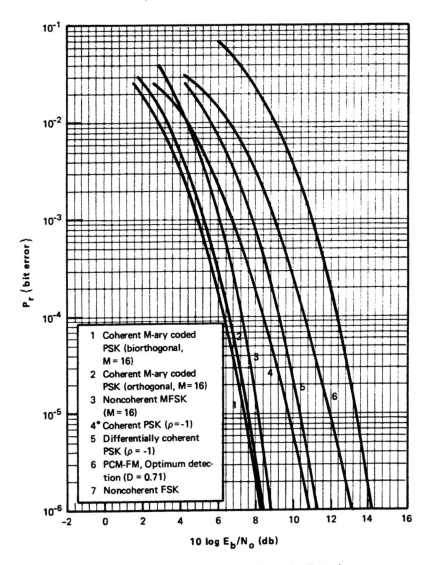

1 Coherent M-ary coded PSK (biorthogonal, M = 16)
2 Coherent M-ary coded PSK (orthogonal, M = 16)
3 Noncoherent MFSK (M = 16)
4 Coherent PSK ($\rho = -1$)
5 Differentially coherent PSK ($\rho = -1$)
6 PCM-FM, Optimum detection (D = 0.71)
7 Noncoherent FSK

— A comparison of M-ary and binary signaling techniques relating signal to noise ratio to the probability of bit error; ex. 10^{-3} means one bit lost or in error per thousand bits.

Digital Modulation Methods	Required Power BER = 10^{-5}	Required Bandwidth
16-ary coded PSK	0 dB (reference)	4 x bit rate
16-ary MFSK	0.8 dB more	4.5 x bit rate
Quadriphase	2.5 dB more	1 x bit rate
Coherent PSK	2.5 dB more	2 x bit rate
PCM/FM	4.6 dB more	2 x bit rate
FSK	6.0 dB more	3 x bit rate

E_b/N_o and the terminal sensitivity factor C/KT_s are related by the bit rate R_b as expressed by:

$$\frac{C}{KT_s} = \frac{C\tau_b}{N_o} \cdot \frac{1}{\tau_b} = \frac{E_b}{N_o} \cdot R_b$$

where $\tau_b = R_b^{-1}$, the bit duration time; $N_o = KT_s$, the noise power density; and $C\tau_b = E_b$.

Courtesy of C.L.Cuccia, W.J.Gill, L.H.Wilson, Philco - Ford.

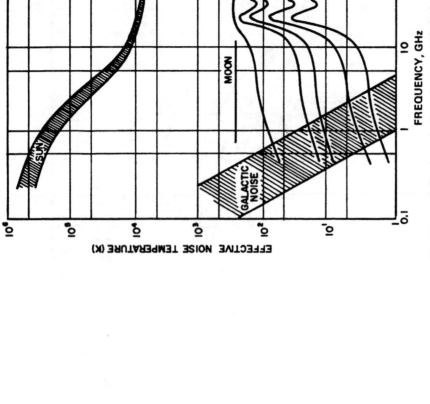

Variation of effective galactic and atmospheric noise temperatures. Degrees on atmospheric noise curves indicate angle above horizon. Minimum noise from these sources is found in the frequency range from 1 to 10 gigahertz.

Courtesy J. W. Thatcher, Deep Space Communications Space/Aeronautics, July 1964.

RADIO STAR	FREQUENCY (MHz)	FLUX DENSITY EQUATION (W/M²/Hz)
CAS-A	40-10,000	$F = 9 \times 10^{-21} f_{MHz}^{-0.82}$
CYG-A	15-40	$F = 7 \times 10^{-22} f_{MHz}^{0.14}$
	400-10,000	$F = 1.16 \times 10^{-20} f_{MHz}^{-0.925}$
	40-400	$F = 3.6 \times 10^{-21} f_{MHz}^{-0.73}$
TAU-A	20-10,000	$F = 6.4 \times 10^{-23} f_{MHz}^{-0.276}$
CEN-A	20-10,000	$F = 3.85 \times 10^{-22} f_{MHz}^{-0.665}$
VIR-A	20-10,000	$F = 6.58 \times 10^{-22} f_{MHz}^{-0.90}$

Flux Density Spectra of Radio Stars

Ultra-Low-Noise TWTA's in 1969

High Dynamic Range with Low Noise Figure and High Power Output

TWTA noise figure performance in the 2 to 18 GHz range with power output as a parameter is:

Frequency (GHz)	Noise Figure		
	P_o = 0 dBm	P_o = 10 dBm	P_o = 20 dBm
2.0 - 4.0	5.5 dB	7.5 dB	10 dB
4.0 - 8.0	6.5 dB	7.5 dB	10 dB
8.0 - 12.0	7.5 dB	8.0 dB	10 dB
12 0 - 18.0	9.5 dB	10.0 dB	12 dB

Extended Bandwidths

Wide bandwidth is an inherent property of the low-noise TWTA. Some typical performance characteristics of double-band amplifiers are

Frequency (GHz)	Noise Figure (dB)	Gain (dB)	Power Output (dBm)
1.0 - 4.0	7.0	25	3
2.0 - 8.0	7.5	25	3
8.0 - 18.0	9.5	25	3
18.0 - 40.0	14.0	25	5

High Power Low-Noise Tubes

High power low-noise tubes have been developed to cover the standard frequency bands.   Typical performance is:

Frequency (GHz)	Power Output (dBm)	Noise Figure (dB)	Gain
2.0 - 4.0	28	12	35
4 0 - 8.0	33	15	40
7.0 - 11.0	33	20	40

Millimeter Low Noise Tubes

State-of-the-art K and Ka-band low-noise TWTA's covering the 18-40 GHz frequency range have also recently been developed. Guaranteed performance characteristics are:

	18.0-26.5 GHz	26.0-40.0 GHz
Noise Figure	11.0 dB	14.0 dB
Gain	25 dB	25 dB
Power	3 dBm	5 dBm
Weight	17 lbs.	25 lbs.
Size	4.75x4. 75x12	5.1x7.5x12.2

Courtesy of J.N.Nelson, Watkins - Johnson Company, Palo Alto, California. Reference: IEEE Convention Digest, March 1969.

147

Modern Low Noise Microwave Technology in 1969

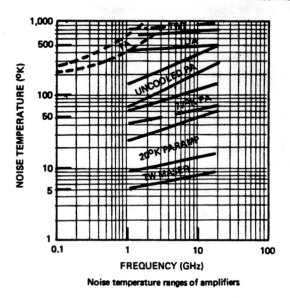

Noise temperature ranges of amplifiers

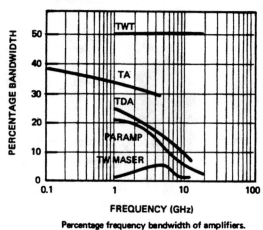

Percentage frequency bandwidth of amplifiers.

Low Noise Transistors and Tunnel Diode Amplifiers in 1969

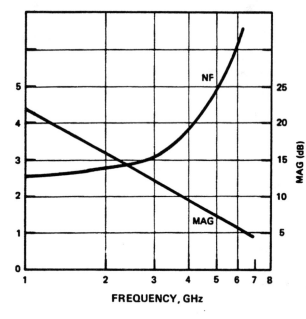

Noise Figure and Power Gain as
a Function of Frequency for Currently
Available Microwave Transistors

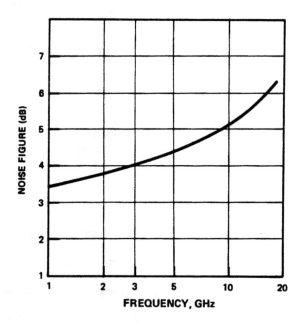

Noise Figure As a Function of
Frequency for Currently Available
Tunnel Diode Amplifiers

Courtesy of Michiyuki Uenohara, Nippon Electric Co., Ltd., Kawasaki City, Japan, and Vladimir G. Gelnovatch,
U.S. Army Electronics Command, Ft. Monmouth, N.J. Reference: IEEE Convention Digest, March 1969.

MICROWAVE TUBES

Performance Characteristics of Principal Frontier Tubes

1969 State of The Art

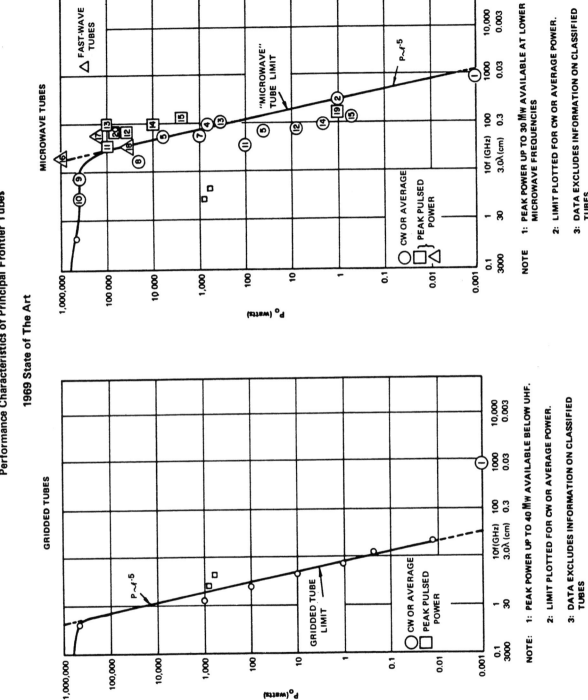

Courtesy of Dr. John Osepchuk

1970 PERFORMANCE CHARACTERISTICS OF PRINCIPAL FRONTIER TUBES

NO.	TUBE TYPE	FREQ. (GHz)	P_o (WATTS)	V (kV)	I_o (A)	EFFICIENCY η (%)	GAIN (dB)	MFG.
1	C004	800	.001	10	0.0033	0.003	--	CSF
2	C010	300	1.0	6	0.083	0.2	--	CSF
3	C020	140	8	6	0.044	3	--	CSF
4	TWT	94	900	32	.280	30	30	Hughes
5	C040	70	40.0	6	0.067	10		CSF
6	819H	55	6kW	42	0.86	16	17	Hughes
7	813H	55	1kW	25	0.3	15	20	Hughes
8	Klystron	18	20kW	23	3.5	25	50	Varian
9	X-3030 (Exper.) (Klystron)	8	400kW	122	10.6	40		Eimac-Varian
	Also VA949AM	7.8	250kW	54	11.0	43	60	Varian
10	Amplitron	3	400kW	--	--	70	--	Raytheon
	Also 5kM1000SG	2.4	475kW					Varian
11	SFD327 Magnetron	35	150kW (Peak)	23	23	25	--	SFD
12	Magnetron	76	40kW (Peak)	16	14	18		Phillips
13	O-BWO WJ-225	100	100kW (Peak) (365W Ave.)	150-200	11	--	--	Watkins-Johnson
14	DX287 (Magnetron)	95	10kW (Peak)	--	--	--	--	Phillips
15	Magnetron	120	2.5kW (Peak)	9	11	2.5		Phillips
16	Ubitron	20	1MW (Peak)	70	230	6	--	G.E.
17	Ubitron	55	150kW (Peak)	70	33 (est.)	6 (est.)	--	G.E.
18	Rippled Beam Amplifier	34	36kW (Peak)	80	6	7	--	MO Valve
19	Electron Cyclotron Maser	140	2 (Peak)	--	--	2	--	Bott
20	Coaxial Magnetrons	70	75kW (Peak)	--	--	--	--	SFD
21	Crossed Field Cyclotron Resonance Maser	38	1kW	14	0.6	10-25	--	Antakov (USSR)
22	VA915A (Twystron)	3.0 (15%) BW	10MW (Peak) (2kW Ave.)	165	165	--	38	Varian
23	L-3775 (Klystron)	UHF	30MW (Peak) (30kW Ave.)	--	--	--	--	Litton
24	SFD303 (Coax Magnetron)	9.4	1MW (Peak) (1kW Ave.)	33	60	50	--	SFD

Courtesy of Dr. John Osepchuk

Microwave Engineers'

I970 State of the Art — Magnetrons

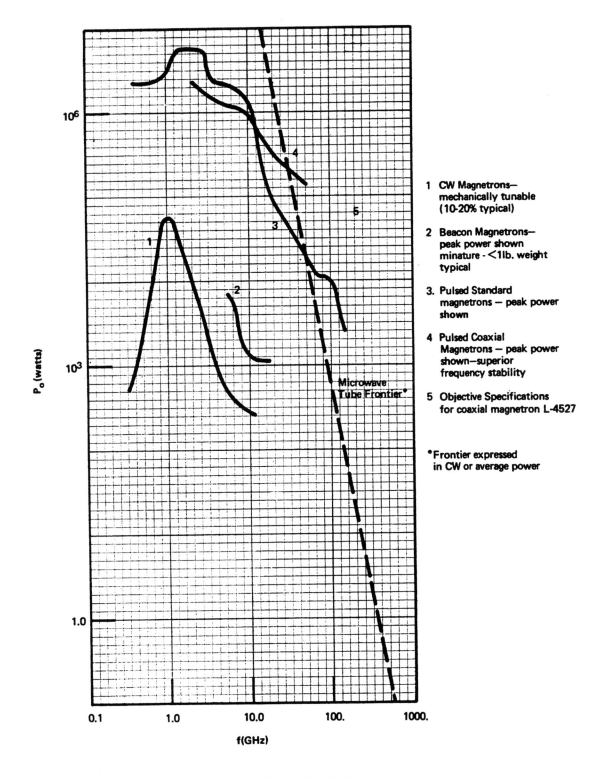

1 CW Magnetrons—
   mechanically tunable
   (10-20% typical)

2 Beacon Magnetrons—
   peak power shown
   minature - <1lb. weight
   typical

3. Pulsed Standard
   magnetrons — peak power
   shown

4 Pulsed Coaxial
   Magnetrons — peak power
   shown—superior
   frequency stability

5 Objective Specifications
   for coaxial magnetron L-4527

*Frontier expressed
 in CW or average power

Courtesy of Dr. John Osepchuk

1970 State of the Art — Power Klystrons

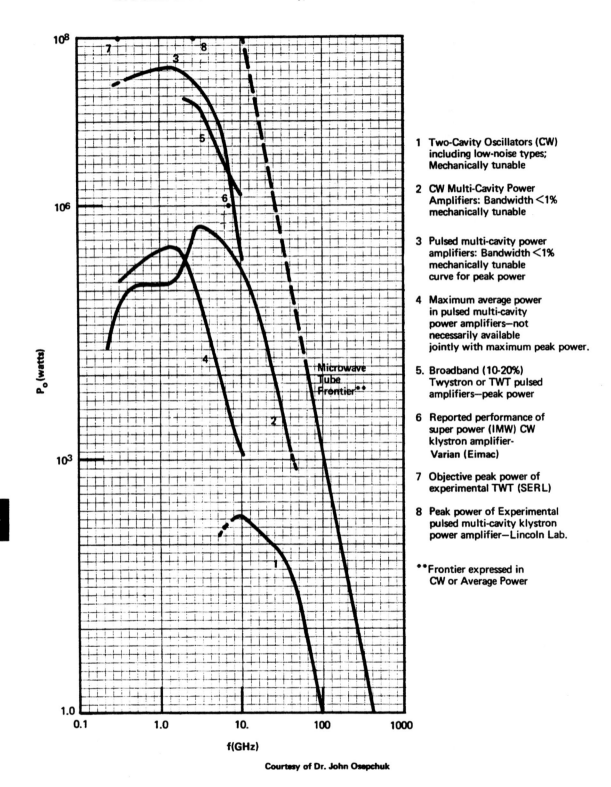

1 Two-Cavity Oscillators (CW) including low-noise types; Mechanically tunable

2 CW Multi-Cavity Power Amplifiers: Bandwidth <1% mechanically tunable

3 Pulsed multi-cavity power amplifiers: Bandwidth <1% mechanically tunable curve for peak power

4 Maximum average power in pulsed multi-cavity power amplifiers—not necessarily available jointly with maximum peak power.

5. Broadband (10-20%) Twystron or TWT pulsed amplifiers—peak power

6 Reported performance of super power (IMW) CW klystron amplifier- Varian (Eimac)

7 Objective peak power of experimental TWT (SERL)

8 Peak power of Experimental pulsed multi-cavity klystron power amplifier—Lincoln Lab.

**Frontier expressed in CW or Average Power

Courtesy of Dr. John Osepchuk

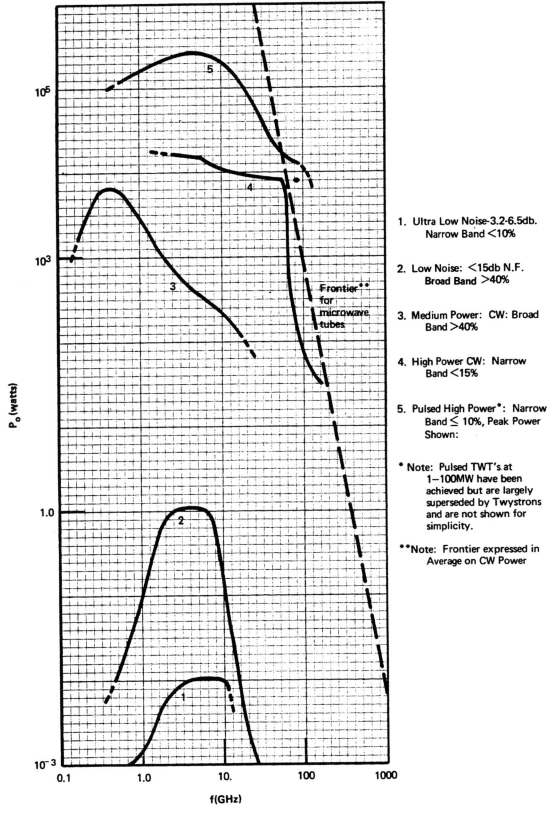

1970 State of the Art — Traveling-Wave Tubes

1. Ultra Low Noise-3.2-6.5db.
   Narrow Band <10%

2. Low Noise: <15db N.F.
   Broad Band >40%

3. Medium Power: CW: Broad
   Band >40%

4. High Power CW: Narrow
   Band <15%

5. Pulsed High Power*: Narrow
   Band ≤ 10%, Peak Power
   Shown:

* Note: Pulsed TWT's at
   1–100MW have been
   achieved but are largely
   superseded by Twystrons
   and are not shown for
   simplicity.

**Note: Frontier expressed in
   Average on CW Power

Courtesy of Dr. John Osepchuk

1970 State of the Art — O-Type BWO's and Reflex Klystrons

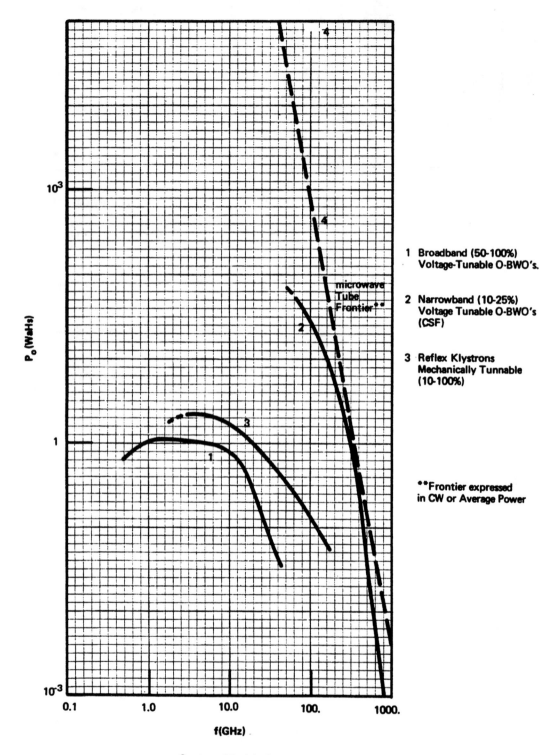

1 Broadband (50-100%)
   Voltage-Tunable O-BWO's.

2 Narrowband (10-25%)
   Voltage Tunable O-BWO's
   (CSF)

3 Reflex Klystrons
   Mechanically Tunnable
   (10-100%)

**Frontier expressed
in CW or Average Power

Courtesy of Dr. John Osepchuk

1970 State of the Art

Crossed Field Amplifiers and Voltage Tunable Oscillators

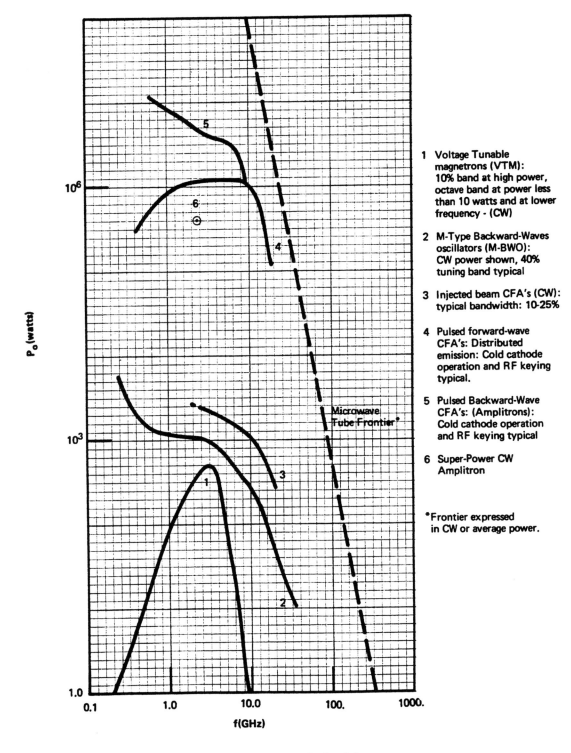

1 Voltage Tunable
   magnetrons (VTM):
   10% band at high power,
   octave band at power less
   than 10 watts and at lower
   frequency - (CW)

2 M-Type Backward-Waves
   oscillators (M-BWO):
   CW power shown, 40%
   tuning band typical

3 Injected beam CFA's (CW):
   typical bandwidth: 10-25%

4 Pulsed forward-wave
   CFA's: Distributed
   emission: Cold cathode
   operation and RF keying
   typical.

5 Pulsed Backward-Wave
   CFA's: (Amplitrons):
   Cold cathode operation
   and RF keying typical

6 Super-Power CW
   Amplitron

*Frontier expressed
 in CW or average power.

Courtesy of Dr. John Osepchuk

157

SOLID STATE

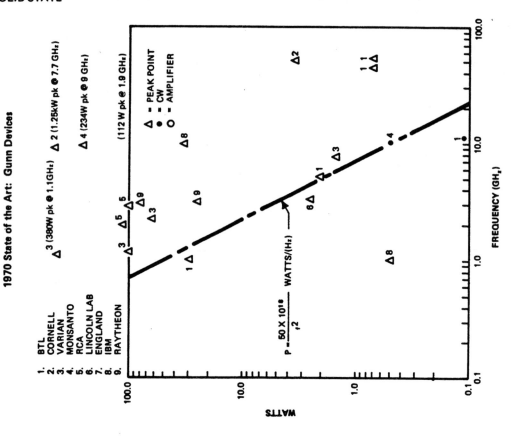

1970 State of the Art: Gunn Devices

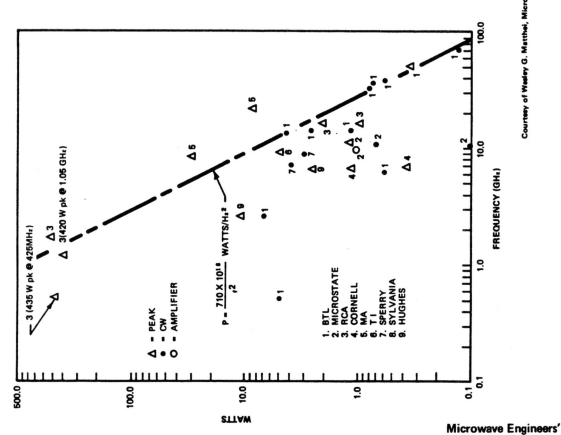

1970 State of the Art: Avalanche (Impatt) Diode

160

Courtesy of Wesley G. Matthei, Micro State Electronics Operation of Raytheon, Murray Hill, N.J.

Microwave Engineers'

1970 State of the Art Results, Power vs. Frequency for EHF Devices

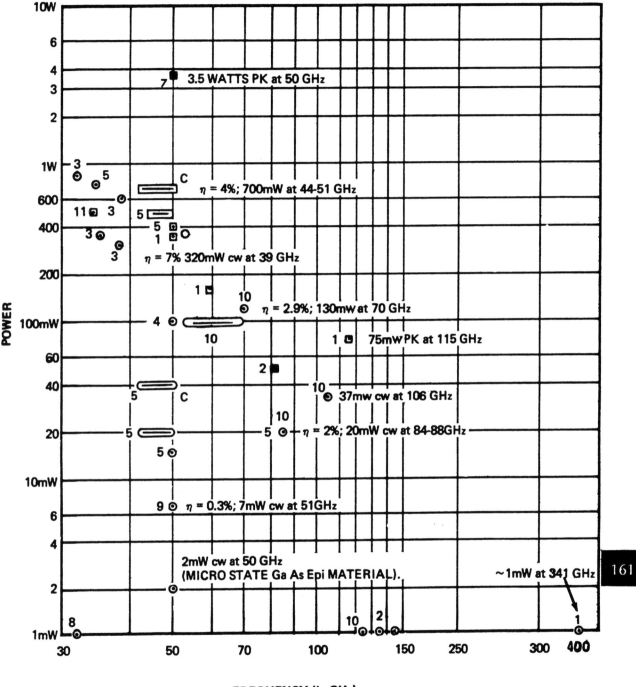

Courtesy of Wesley G. Matthei, Micro State Electronics Operation of Raytheon, Murry Hill, N.J.

1970 State of the Art: GUNN, LSA and Hybrid Devices Using Bulk GaAs

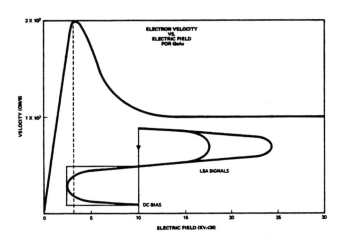

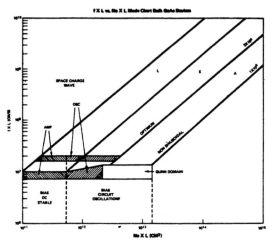

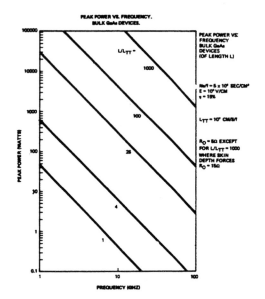

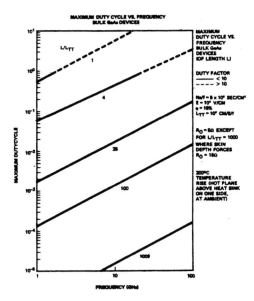

Courtesy of Lester F. Eastman, Cornell University, Ithaca, N.Y.

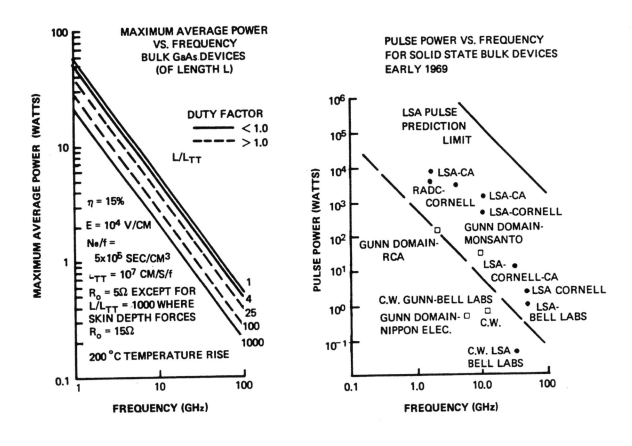

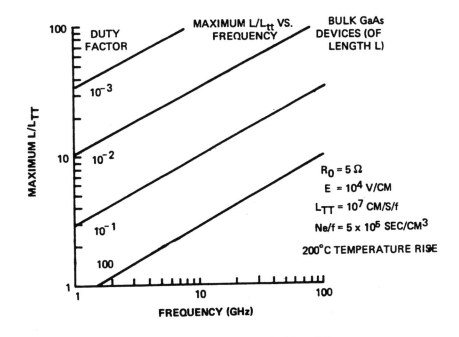

Courtesy of Lester F. Eastman, Cornell University, Ithaca, N.Y.

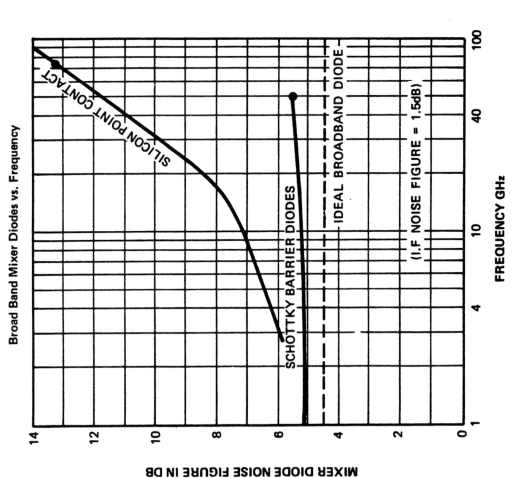

SOLID STATE

1970 State of The Art: Effective Input Noise
Temperature of Parametric Amplifiers vs. Frequency

TUNNEL DIODES

TRAVELING-WAVE TUBES

CONVENTIONAL PARAMP

"BEST" PARAMP-UNCOOLED

"BEST" PARAMP-COOLED TO N²

"BEST" PARAMP-COOLED TO 20°K

TRAVELING-WAVE MASERS

NOISE TEMPERATURE IN K

FREQUENCY IN GHz

1970 State of The Art: System Noise Figure of
Broad Band Mixer Diodes vs. Frequency

SILICON POINT CONTACT

SCHOTTKY BARRIER DIODES

IDEAL BROADBAND DIODE

(I.F NOISE FIGURE = 1.5dB)

MIXER DIODE NOISE FIGURE IN DB

FREQUENCY GHz

164

1970 State of the Art of the Noise Measure
of Impatt and Bulk GaAs Devices

		SMALL SIGNAL AMPLIFIER	LARGE SIGNAL OSCILLATOR
IMPATT	SILICON	40dB	55dB
	GERMANIUM	30dB	40dB
	Ga As	25dB	35dB *

		SMALL SIGNAL AMPLIFIER	LARGE SIGNAL OSCILLATOR
BULK Ga As	DOMAIN	N/A	25dB-45dB*
	IDEAL LSA	N/A	EST. 10dB *
	ACCUMULATION	20dB	36dB *
	THIM TYPE II	22dB	N/A

* FLICKER NOISE NOT CONSIDERED

Courtesy of J.Josenhans, Bell Telephone Laboratories, Murray Hill, N.J.

1970 Dynamic Range Comparison Chart for Microwave Detectors

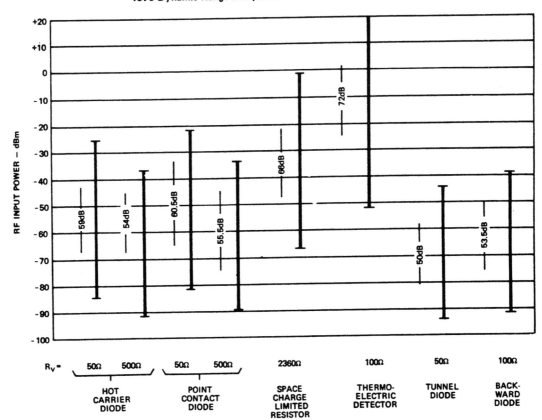

Courtesy of A.M.Cowley and H.O.Sorensen, Hewlett - Packard Co., Palo Alto, California.

165

1970 State of the Art: Tunnel Diode Amplifiers and Transistors, Noise Figure vs. Frequency

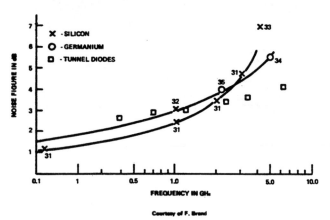

Courtesy of F. Brand

REFERENCES

1. Prager, H. J., et al., "High Efficiency Silicon Avalanche Diodes at Ultra-High Frequencies", Proc. IEEE, Vol. 55, April 1967, p.p. 234-235

2. Prager, H. J., et al., "Anomalous Avalanche Diode", 25th Conference on Electron Devices Research, McGill University, Montreal, Canada, June 21-23, 1967

3. Dow, D. G., "Two Gunn Holdups Slow Microwave Oscillators", Electronics, November 13, 1967, p.p. 129-130

4. Cornblest, S., "Microwave Solid State Activity in the United Kingdom", Microwave Journal, February 1968, p.p.45

5. Kilpatrick, S., "Microwave News", Microwaves, June 1967, p.p. 10-12

6. Eastman, L. F., Informal comments at "ICAMEBS" Conference on Active Microwave Effects in Bulk Semiconductors, New York, N. Y., January 1968

7. Cover Feature, Microwaves, October 1967, p.p. 76

8. Kennedy, W. K., Jr., et al., "LSA Operation of large Volume Bulk GaAs Samples", IEEE Transactions on Electron Devices, Vol. ED-14, Sept. 1967, p.p. 500-504

9. CA5X2-E, Cayuga Associates

10. Gilden, M., and Moroney, W., "High Power Pulsed Avalanche Diode Oscillators for Microwave Frequencies", Proc. IEEE, Vol. 55, July 1967, p.p. 1227-1228

11. Kennedy, W. K., Eastman, L. F., "High Power Pulsed Microwave Generation in Gallium Arsinide", Proc. IEEE, Vol. 55, No. 3, March 1967, p.p. 434-435

12. Swan, C. B., "Improved Performance of Silicon Avalanche Oscillators Mounted on Diamond Heat Sinks, Proc. IEEE, Vol. 55, September 1967, p.p. 1617-1618.

13. Dunn, C. N., and H. M. Olson, "Germanium Epitaxial Mesa Diodes for a 6 GHz, ½ Watt CW Avalanche Oscillator", 1967 IEEE International Electron Devices Meeting, October 1967, Washington, D. C.

14. Bomac B 10, Varian Associates, Bomac Division

15. Swan, C. B., et al., "Continuous Oscillation at Millimeter Wavelength With Silicon Avalanche Diodes", Proc. IEEE, Vol. 55, October 1967, p.p. 1747-1748

16. Eastman, L. F., "Semi-Conductor Microwave Power Generation", Solid State Microwave 1968 Tuesday Evening Lecture Series, Sponsored by Washington,

D. C. Section IEEE Groups on Microwave Theory and Techniques, Feb. 27, 1968

17. Misawa, T., "CW Millimeter-Wave IMPATT Diodes with Nearly Abrupt Junctions, Proc. IEEE, Vol. 56, February 1968, p.p. 234-235

18. Copeland, J. A., "CW Operation of LSA Oscillator Diodes", Bell System Technical Journal, January 1967, p.p. 284-287

19. Murakami, H., et al., "Electrical Performance of GaAs Epitaxial Gunn Effect Oscillators", IEEE Trans. On Electron Devices, Vol. ED-14, September 1967, p.p. 611-612

20. Griffin, H., and Segal, S., "Final Progress Report for UHF Duplexers", U. S. Navy Contract NObsr 89105, May 1964

21. Barber, M. R., Sodomsky, K. F., and Zacharias, A. "Microwave Semiconductor Devices and Their Circuit Applications", Ed. by H. A. Watson, Chap. 10

22. Mortenson, K. E. and White, J. F., "X Band Room Temperature Bulk Effect High Power Limiter", ISSCC Digest of Technical Papers 1967

23. Ryder, R. M., Brown, N. J., and Forest, R. G., "Microwave Diode Control Devices, Part I - Diodes and Passive Limiters", Microwave Journal, February 1968, p.p. 57-64

24. PT 6650 (TRW)

25. 2n6178 (TRW)

26. Sterzer, F., "Microwave Solid-State Power Sources", Microwave Journal, Feb 1968, p.p. 67-74

27. S1060 (United Aircraft Corp.), Electronics Component Div.)

28. Carley, D., RCA Summerville, Personal Communication

29. Webster, R., Informal comments, 1968 G-MTT Symposium, Detroit, Michigan

30. 2N4976 (TRW)

31. L194 (Texas Instruments)

32. KD5201 (KMC Semiconductor Corp.)

33. Bell Telephone Laboratories, In-house (Production)

34. TIXM103 Texas Instruments

35. TIXM105-106 Texas Instruments

36. Mortenson, K. E., Private communication

1970 State of the Art: Semiconductor Limiting Devices Reliable Power Handling Capability vs. Frequency

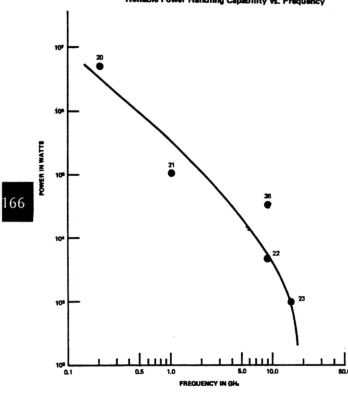

Courtesy of F. Brand

1970 State of the Art Microwave Transistors Power Source or Amplifiers

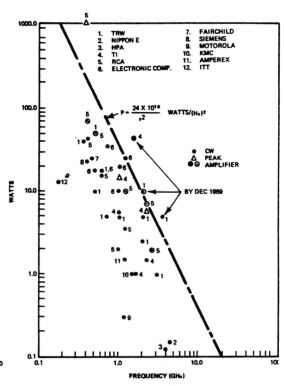

Courtesy of Wesley G. Matthei, Micro State Electronics Operation of Raytheon, Murray Hill, N.J.

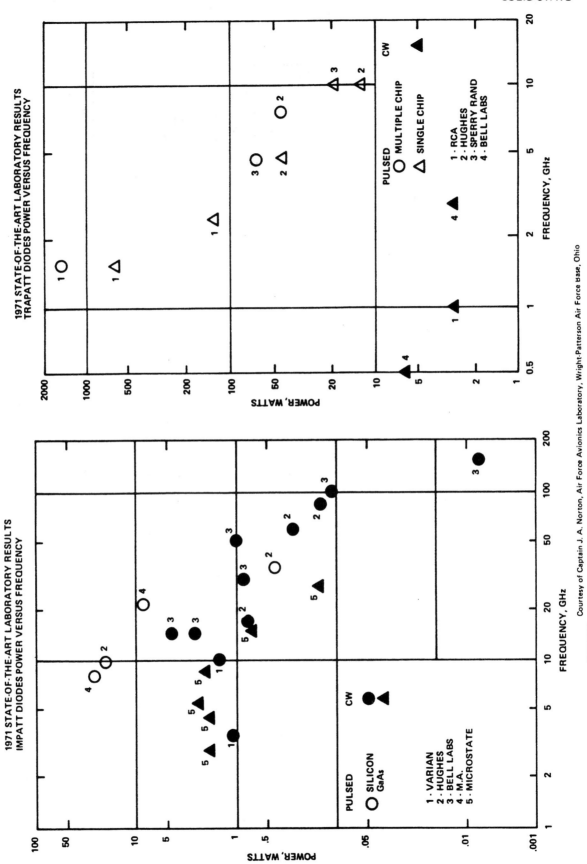

1971 STATE-OF-THE-ART LABORATORY RESULTS
TRAPATT DIODES POWER VERSUS FREQUENCY

1971 STATE-OF-THE-ART LABORATORY RESULTS
IMPATT DIODES POWER VERSUS FREQUENCY

Courtesy of Captain J. A. Norton, Air Force Avionics Laboratory, Wright-Patterson Air Force Base, Ohio

167

168

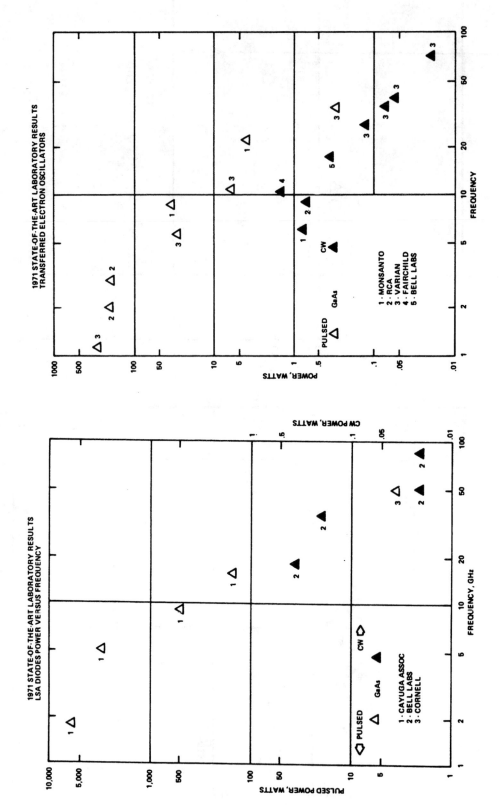

Courtesy of Captain J. A. Norton, Air Force Avionics Laboratory, Wright-Patterson Air Force Base, Ohio

Dissipative Effect of Power Addition For Power Amplifiers

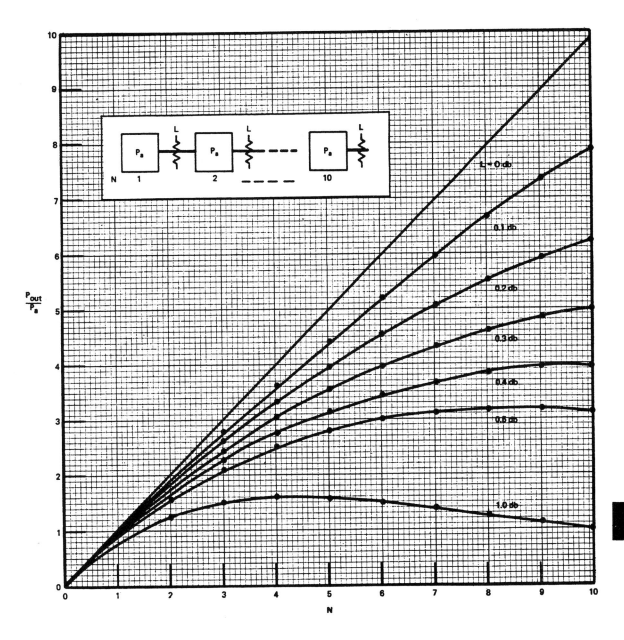

Courtesy of John Sie and Chung Kim, Micro State Electronics, Operation of Raytheon, Murray Hill, N.J.

RATIO OF JUNCTION CAPACITANCE AT REVERSE BIAS VOLTAGE (v)
TO JUNCTION CAPACITANCE AT 6 VOLTS REVERSE BIAS
VERSUS APPLIED REVERSE Vo

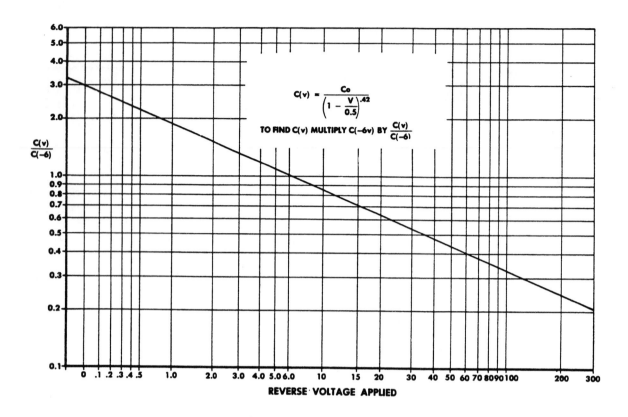

$$C(v) = \frac{C_o}{\left(1 - \frac{V}{0.5}\right)^{.42}}$$

TO FIND C(v) MULTIPLY C(-6v) BY $\frac{C(v)}{C(-6)}$

Courtesy of Sylvania Electric Products, Inc., from the Sylvania Microwave Diode Product Guide.

ELECTRONIC CONSTANTS FOR TYPICAL SEMICONDUCTORS*

Semiconductor	Band Gap (Electron Volts)	Hole Mobility μ_p (cm²/volt-sec)	Electron Mobility μ_n (cm²/volt-sec)	Dielectric Constant (K)
Ge	0.67	1900	3900	16
Si	1.11	500	1500	12
C	6.7	1200	1800	—
GaP	2.25	>20	>100	—
AlSb	1.52	460	460	11.5
GaAs	1.4	450	9000	13.5
InP	1.25	150	4800	10.6
GaSb	0.8	900	4500	15.2
InAs	0.35	450	33000	11.5
InSb	0.18	3000	85000	16.5

* These data have been compiled from various references in the literature and may be subject to correction due to refinement in measurement.

NOMOGRAPH FOR CALCULATION OF SERIES RESISTANCE
OF VARACTOR DIODES

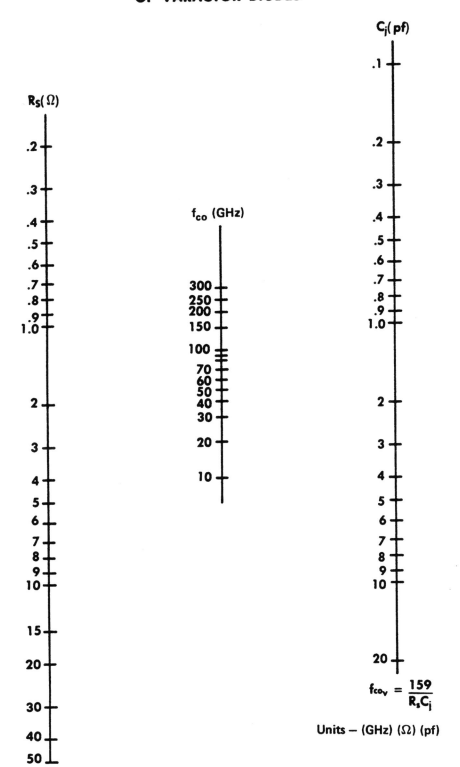

$$f_{co_v} = \frac{159}{R_s C_j}$$

Units — (GHz) (Ω) (pf)

Courtesy of Sylvania Electric Products, Inc.

171

TUNNEL DIODE MEASUREMENTS
RESISTIVE CUT-OFF FREQUENCY f_{∞}

Chart for Calculation of Cut-off Frequency of Tunnel Diodes.

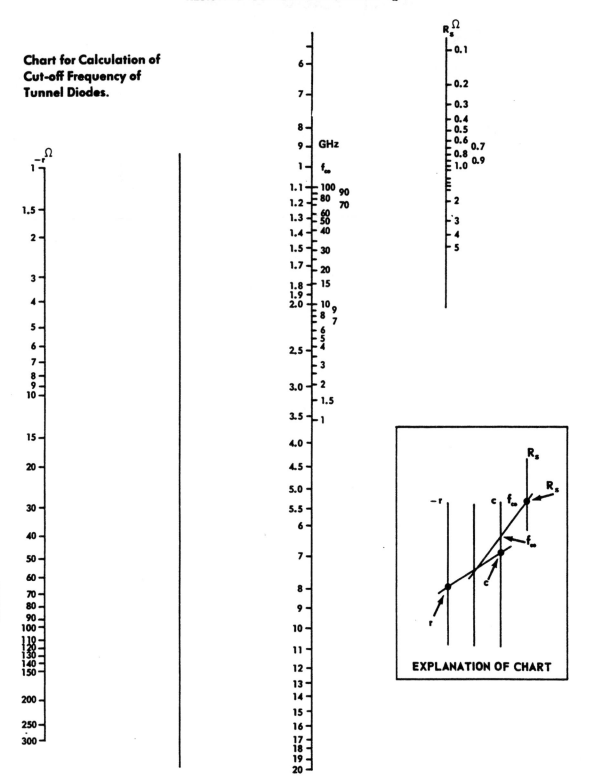

EXPLANATION OF CHART

Courtesy of Sylvania Electric Products, Inc., from the Sylvania Microwave Diode Product Guide.

TUNNEL DIODE FIGURE OF MERIT

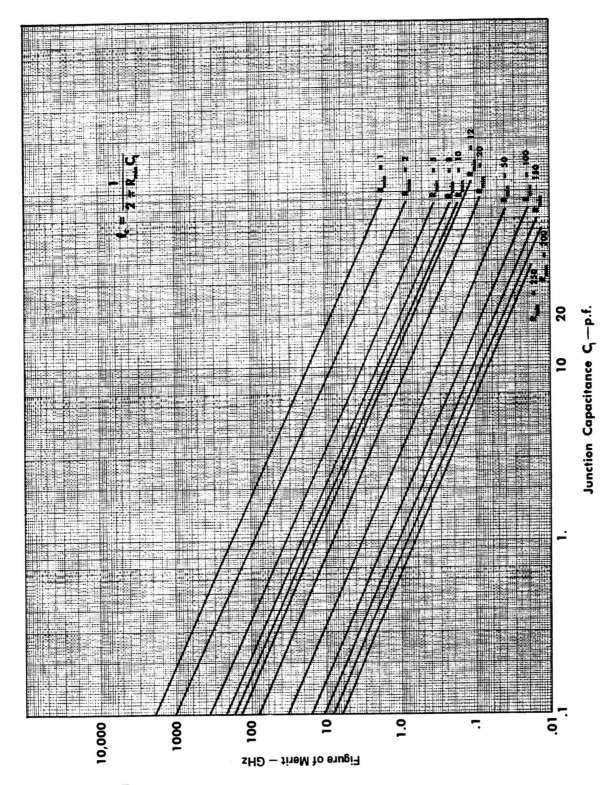

PIN DIODE CIRCUIT DESIGN CURVES

INSERTION LOSS vs R $_{SERIES}$ FOR A DIODE IN SERIES WITH A 50 OHM LOAD

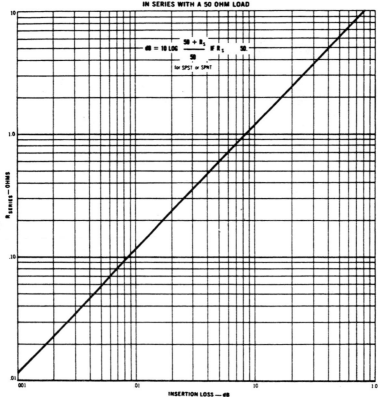

PIN DIODE CIRCUIT DESIGN CURVES

INSERTION LOSS vs R $_{SHUNT}$

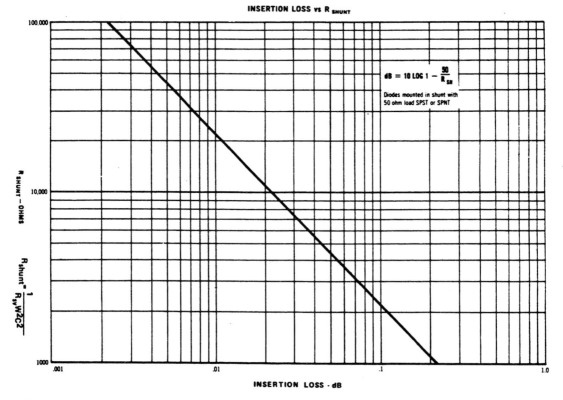

Courtesy of Unitrode Corporation, Watertown, Massachusetts

Microwave Engineers'

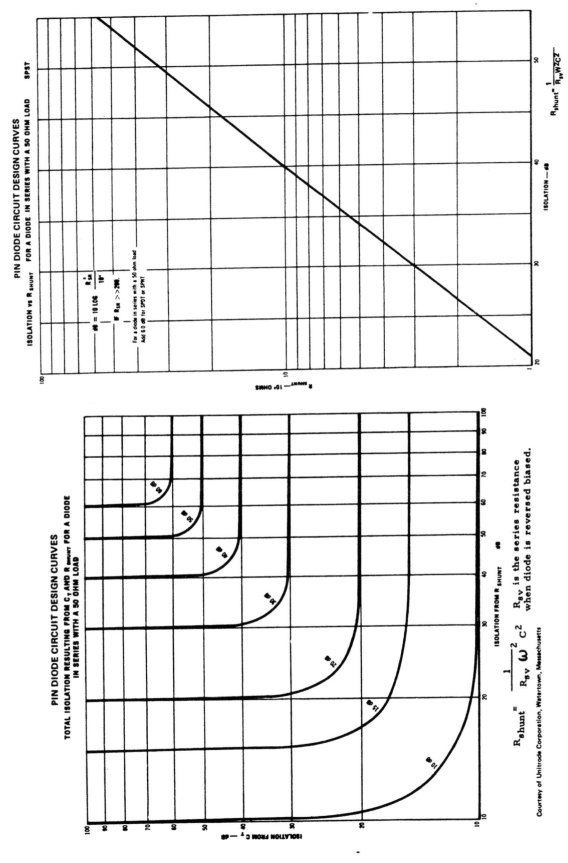

175

PIN DIODE CIRCUIT DESIGN CURVES

ISOLATION vs BANDWIDTH FOR VARIOUS VALUES OF C_T FOR SPST

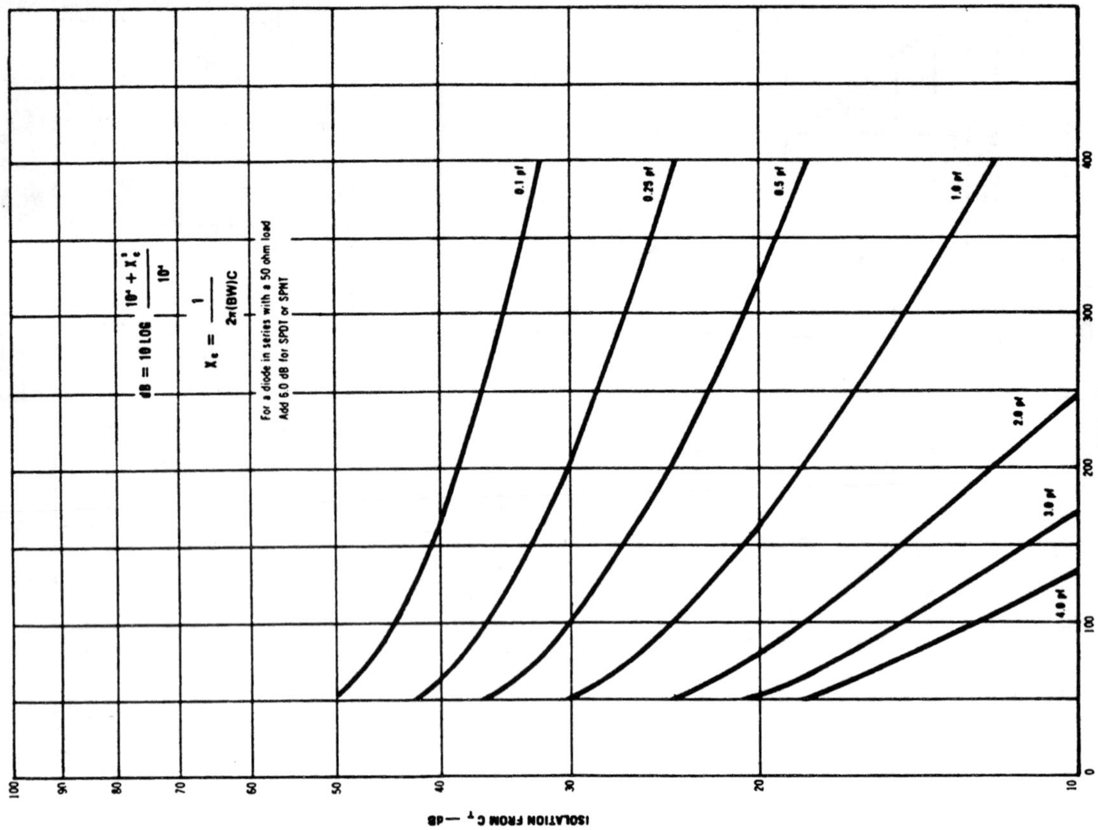

$dB = 10 \log \frac{10^4 + X_c^2}{10^4}$
$X_c = \frac{1}{2\pi(BW)C}$
For a diode in series with a 50 ohm load
Add 6.0 dB for SPDT or SPNT
0.1 pf
0.25 pf
0.5 pf
1.0 pf
2.0 pf
3.0 pf
4.0 pf
ISOLATION FROM C_T — dB
BANDWIDTH — MHz
100
90
80
70
60
50
40
30
20
10
0
100
200
300
400

PIN DIODE CIRCUIT DESIGN CURVES

ISOLATION vs SERIES RESISTANCE (R_s)

$dB = 10 \log \frac{50^2}{4R_s^2}$
Diode in shunt with 50 ohm load in SPST
Add 6.0 dB for SPDT or SPNT
SERIES RESISTANCE — R_s — OHMS
ISOLATION — dB
10
1.0
.10
10
20
30
40
50
60

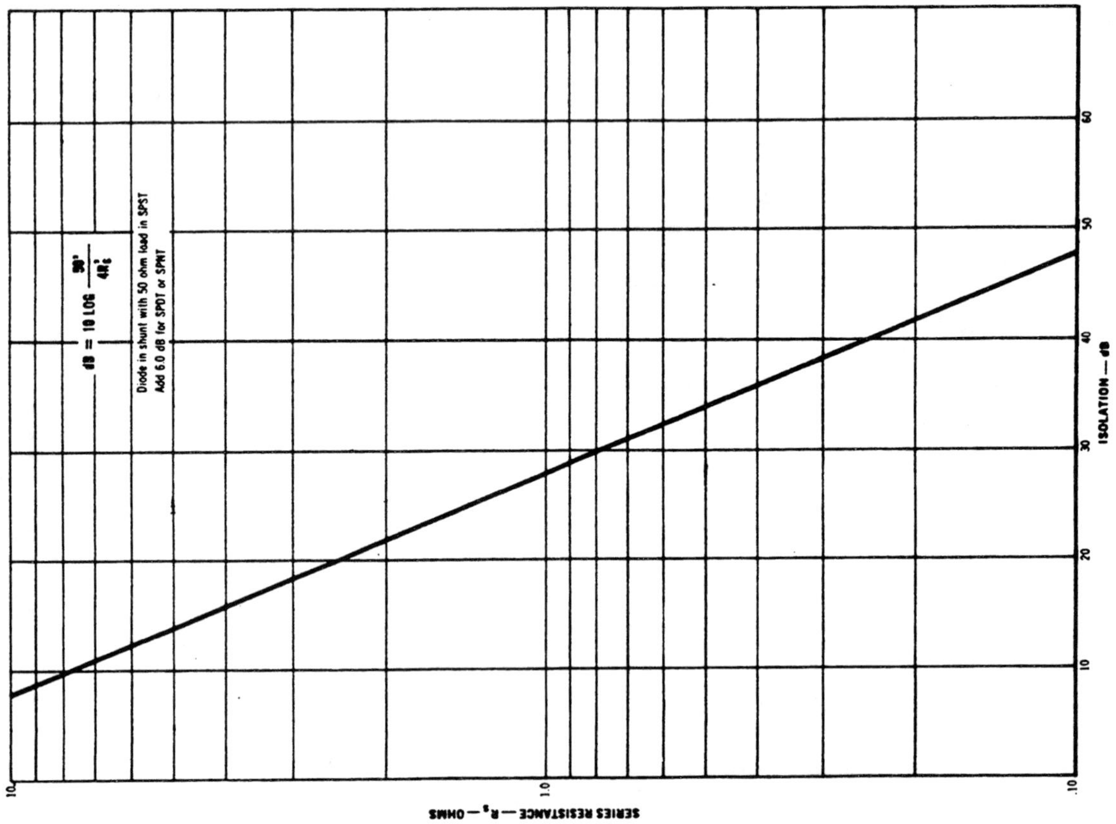

Courtesy of Unitrode Corporation, Watertown, Massachusetts

PIN DIODE POWER HANDLING DESIGN CURVES

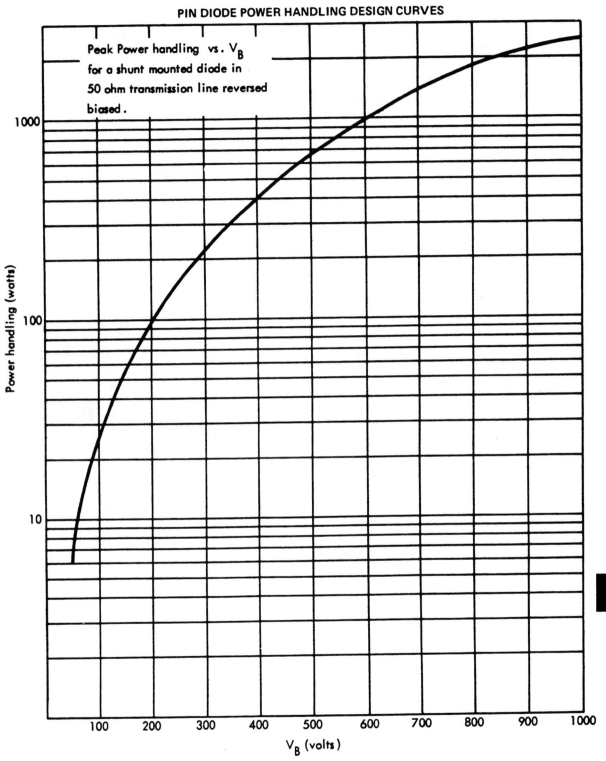

Peak Power handling vs. V_B for a shunt mounted diode in 50 ohm transmission line reversed biased.

Power handling (watts)

V_B (volts)

Courtesy of Sylvania Electric Products Inc., Woburn, Massachusetts

177

X-BAND

**VOLTAGE OUTPUT VS. POWER INPUT
OF POINT CONTACT DIODES**

X-BAND

**VOLTAGE OUTPUT VS. POWER INPUT
OF SCHOTTKY BARRIER DIODES**

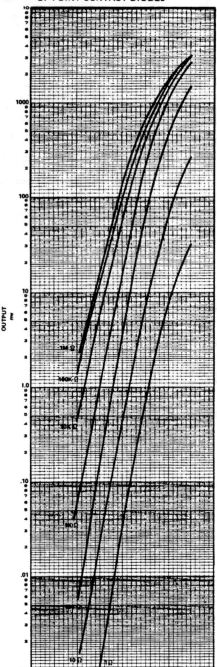

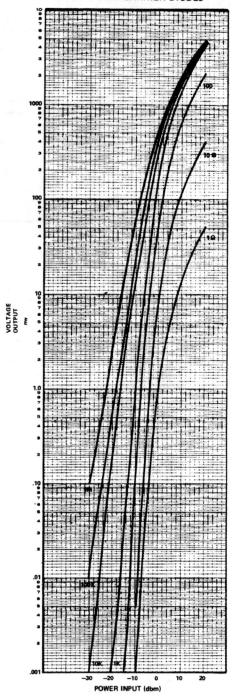

Courtesy of Sylvania Electric Products Inc., Woburn, Massachusetts

MISCELLANEOUS

IF AMPLIFIER DESIGN INFORMATION

1. The impedance and transfer characteristics of active devices (such as bipolar transistors, field effect transistors, and tubes), when operating in their linear range, can be expressed in terms of small signal parameters. Commonly used parameters are the s, h, y and z parameters. S parameters are usually used at microwave frequencies, y at intermediate frequencies and h at audio frequencies. Figure 1 shows the derivation of the y parameters from a two port network.

Figure 1

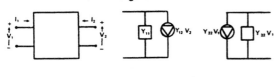

$$Y_{11} = \frac{I_1}{V_1}, \quad V_2 = 0 \qquad Y_{12} = \frac{-I_1}{V_2}, \quad V_1 = 0$$

$$Y_{22} = \frac{I_2}{V_2}, \quad V_1 = 0 \qquad Y_{21} = \frac{I_2}{V_1}, \quad V_2 = 0$$

Parameters can be transformed from one form to another as shown in Figure 2*.

Figure 2

s-parameters in terms of h-, y-, and z-parameters	h-, y-, and z-parameters in terms of s-parameters
$s_{11} = \dfrac{(Z_{11}-1)(Z_{22}+1) - Z_{12}Z_{21}}{(Z_{11}+1)(Z_{22}+1) - Z_{12}Z_{21}}$	$Z_{11} = \dfrac{(1+s_{11})(1-s_{22}) + s_{12}s_{21}}{(1-s_{11})(1-s_{22}) - s_{12}s_{21}}$
$s_{12} = \dfrac{2Z_{12}}{(Z_{11}+1)(Z_{22}+1) - Z_{12}Z_{21}}$	$Z_{12} = \dfrac{2s_{12}}{(1-s_{11})(1-s_{22}) - s_{12}s_{21}}$
$s_{21} = \dfrac{2Z_{21}}{(Z_{11}+1)(Z_{22}+1) - Z_{12}Z_{21}}$	$Z_{21} = \dfrac{2s_{21}}{(1-s_{11})(1-s_{22}) - s_{12}s_{21}}$
$s_{22} = \dfrac{(Z_{11}+1)(Z_{22}-1) - Z_{12}Z_{21}}{(Z_{11}+1)(Z_{22}+1) - Z_{12}Z_{21}}$	$Z_{22} = \dfrac{(1+s_{22})(1-s_{11}) + s_{12}s_{21}}{(1-s_{11})(1-s_{22}) - s_{12}s_{21}}$
$s_{11} = \dfrac{(1-y_{11})(1+y_{22}) + y_{12}y_{21}}{(1+y_{11})(1+y_{22}) - y_{12}y_{21}}$	$y_{11} = \dfrac{(1+s_{22})(1-s_{11}) + s_{12}s_{21}}{(1+s_{11})(1+s_{22}) - s_{12}s_{21}}$
$s_{12} = \dfrac{-2y_{12}}{(1+y_{11})(1+y_{22}) - y_{12}y_{21}}$	$y_{12} = \dfrac{-2s_{12}}{(1+s_{11})(1+s_{22}) - s_{12}s_{21}}$
$s_{21} = \dfrac{-2y_{21}}{(1+y_{11})(1+y_{22}) - y_{12}y_{21}}$	$y_{21} = \dfrac{-2s_{21}}{(1+s_{11})(1+s_{22}) - s_{12}s_{21}}$
$s_{22} = \dfrac{(1+y_{11})(1-y_{22}) + y_{12}y_{21}}{(1+y_{11})(1+y_{22}) - y_{12}y_{21}}$	$y_{22} = \dfrac{(1+s_{11})(1-s_{22}) + s_{12}s_{21}}{(1+s_{22})(1+s_{11}) - s_{12}s_{21}}$
$s_{11} = \dfrac{(h_{11}-1)(h_{22}+1) - h_{12}h_{21}}{(h_{11}+1)(h_{22}+1) - h_{12}h_{21}}$	$h_{11} = \dfrac{(1+s_{11})(1+s_{22}) - s_{12}s_{21}}{(1-s_{11})(1+s_{22}) + s_{12}s_{21}}$
$s_{12} = \dfrac{2h_{12}}{(h_{11}+1)(h_{22}+1) - h_{12}h_{21}}$	$h_{12} = \dfrac{2s_{12}}{(1-s_{11})(1+s_{22}) + s_{12}s_{21}}$
$s_{21} = \dfrac{-2h_{21}}{(h_{11}+1)(h_{22}+1) - h_{12}h_{21}}$	$h_{21} = \dfrac{-2s_{21}}{(1-s_{11})(1+s_{22}) + s_{12}s_{21}}$
$s_{22} = \dfrac{(1+h_{11})(1-h_{22}) + h_{12}h_{21}}{(h_{11}+1)(h_{22}+1) - h_{12}h_{21}}$	$h_{22} = \dfrac{(1-s_{22})(1-s_{11}) - s_{12}s_{21}}{(1-s_{11})(1+s_{22}) + s_{12}s_{21}}$

*Obtained from Hewlett Packard's "S" Parameter Article Permission not requested.

2. Stable gain of the device is determined by:

$$A_s = \frac{|Y_{21}|}{S|Y_{12}|}$$

where S is the stability factor (usually taken as a value of 4). With a value of S less than unity oscillations will usually occur. A_s is the stable power gain expressed as a ratio.

3. The bandwidth reduction, as a function of the number of cascaded tuned circuits, is shown in Table 1.

TABLE I

Number of Cascaded Stages	Reduction in Bandwidth
1	1.0
2	0.644
3	0.510
4	0.435
5	0.368
6	0.350
7	0.323
8	0.301
9	0.283
10	0.268
15	0.217
20	0.187

TABLE II

FORMULAS RELATING EQUIVALENT SERIES AND PARALLEL COMPONENTS

$$Q = \frac{X_s}{R_s} = \frac{WL_s}{R_s} = \frac{1}{WC_sR_s} = \frac{R_p}{X_p} = \frac{R_p}{WL_p} = R_pWC_p$$

General Formulas	Formulas for Q greater than 10	Formulas for Q less than 0.1
$R_s = \dfrac{R_p}{1+Q^2}$	$R_s = \dfrac{R_p}{Q^2}$	$R_s = R_p$
$X_s = X_p \dfrac{Q^2}{1+Q^2}$	$X_s = X_p$	$X_s = X_pQ^2$
$L_s = L_p \dfrac{Q^2}{1+Q^2}$	$L_s = L_p$	$L_s = L_pQ^2$
$C_s = C_p \dfrac{1+Q^2}{Q^2}$	$C_s = C_p$	$C_s = \dfrac{C_p}{Q^2}$
$R_p = R_s(1+Q^2)$	$R_p = R_sQ^2$	$R_p = R_s$
$X_p = X_s \dfrac{1+Q^2}{Q^2}$	$X_p = X_s$	$X_p = \dfrac{X_s}{Q^2}$
$L_p = L_s \dfrac{1+Q^2}{Q^2}$	$L_p = L_{ss}$	$L_p = \dfrac{L_s}{Q^2}$
$C_p = C_s \dfrac{Q^2}{1+Q^2}$	$C_p = C_s$	$C_p = C_sQ^2$

4. The number of stages required is determined as follows:

$$N = \frac{A \text{ total}}{A_s}$$

Where N is the total number of stages, A is the total gain, and A_s is the stable gain per stage.

Courtesy of Kevin Redmond, LEL Division of Varian, Copiaque, New York

180

5. The bandwidth of a single stage is determined as follows:

$$B_1 = \frac{B_t}{f}$$

Where B_1 is the bandwidth of one stage, B_t is the bandwidth of the total amplifier and f is the factor obtained from Table 1 for N + 1 stages.

6. The bandwidth and Q relationship of a single tuned circuit is as follows:

$$BW = \frac{f_o}{Q} = \frac{1}{2\pi R_p C} \quad , \quad f_o = \frac{1}{2\pi \sqrt{LC}}$$

Where BW is the bandwidth, f_o is the center frequency and Q is the quality factor. R_p is the parallel resistance of the tuned circuit, C is the capacitance and L is the inductance.

7. Useful relationships of Q to the tuned circuit parameters are given in Table II.

8. The relationship of bandwidth and phase, for a single tuned circuit, is given in Table III.

TABLE III

BANDWIDTH AND PHASE RELATIONSHIPS					
BW	% DOWN	REL BW	PHASE ANG	LINPHASE ANGLE	FROM LIN
db					
0.25	.96	25%	15°	15°	0°
0.5	.94	35%	25	26	1
1.0	.89	50%	27.5	30	2.5
2.0	.79	70%	35.0	40	5.0
3.0	.70	100%	45.0	55	10.0
6.0	.50	170%	57.5	95	37.5

9. Third order distortion is given by:

$$P_c = \frac{3P_d - 15 - P_3}{2}$$

Where P_c is the 1db compression point, P_3 is the third order compression level in dBM and P_d' is the desired signal level in dBM.

10. Second order distortion is given by :

$$P_c = 2P_d - 15 - P_3$$

11. Group delay or envelope delay describes the phase slope and can be calculated from the phase characteristic as follows:

$$T_e = \frac{-d\theta}{dw}$$

Where T_e is the envelope delay, θ is the phase characteristic as a function of w.

12. Noise figure of a multistage amplifier is given by the following equation:

$$F_t = F_1 + \frac{F_2 - 1}{G_1} + \frac{F_3 - 1}{G_1 G_2} + \ldots$$

Where F_t is the total noise figure, F_1 is the first stage noise figure, F_2 is the second stage noise figure and G_1 is the first stage gain.

In a transistor, the noise factor is given by:

$$F = 1 + \frac{r_b 1}{R_g} + \frac{re}{2R_g} + \left(\frac{R_g + re + r_b}{2R_g \, re \, h_{fb}} \right)$$

$$\left\{ \left[1 + f/_{fa} \right] \quad \left[\frac{h_{FB}}{h_{fb}} \left(1 + \frac{I_{CBO}}{I_C} \right) - h_f 13 \right] \right.$$

13. Johnson noise of a resistor is given by:

$$E_n^2 = 4KTBR$$

Where

K = Baltzmann's constant
= 1.372×10^{-23} joules per degree Kelvin

T = Temperature of resistor in degrees Kelvin

B = System bandwidth in Hz

R = Value of the resistant in ohms.

14. At room temperature the noise power available from a resistor with a 1 MHz passband is -114dBM.

15. Current commercially available noise figures as a function of frequency:

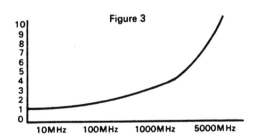

16. The practical efficiency of a class A linear transistor stage is typically less than 10%.

17. A typical transistor stage with a bias network is shown in figure 4.

TYPICAL TRANSISTOR STAGE

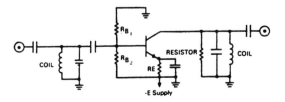

TYPICAL TRANSISTOR STAGE

Figure 4

Courtesy of Kevin Redmond, LEL Division of Varian, Copiaque, New York

181

18. Thermal stability is given by the following equation. Thermal stability is an indication of the bias stability of the circuit with temperature. No stability is attained when the stability factor is equal to B. Excellent stability has been attained when S_t is low (1 to 5).

$$S_t = \frac{1 + \dfrac{R_E}{R_{B_1}} \left[1 + \dfrac{R_{B_1}}{R_{B_2}} \right]}{(1 - a) + \dfrac{R_E}{R_{B_1}} \left[1 + \dfrac{R_{B_1}}{R_{B_2}} \right]}$$

19. The bandwidth rise time products are given as follows:

$$BW\ R_t = .35 \quad \text{(for video)}$$

$$BW\ R_t = .7 \quad \text{(for IF)}$$

Logarithmic amplifiers provide an output signal which is a logarithmic function of the input signal. Usually the output is at video while the input is at IF. The advantage of a logarithmic amplifier is a wide dynamic range without AGC or in effect instantaneous AGC. A typical logarithmic amplifier characteristic is given in Figure 5.

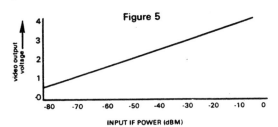

Figure 5

20. The basic block diagrams of a linear and a logarithmic amplifier are shown in Figure 6.

LINEAR AMPLIFIER CONFIGURATION

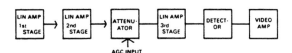

LOGARITHMIC AMPLIFIER CONFIGURATION

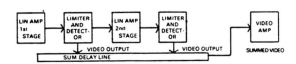

Figure 6

In the successive detection logarithmic amplifier each stage is usually a linear stage followed by a limiter and detector. The stage acts as a linear amplifier until limited. The output of the stage is detected and added to the sum line.

21. The accuracy of a logarithmic amplifier is a function of the number of stages, as shown in Table IV.

TABLE IV

STAGE GAIN	LOG ACCURACY
6dB	.25 dB
8dB	.50 dB
10dB	.75 dB
12dB	1.00 dB

22. The dynamic range of a logarithmic amplifier is given by the following equation:

$$R = (N + 1) G$$

Where R is the dynamic range in dB, N is the number of stages and G is the gain per stage in dB.

23. The apparent increase in rise time and decrease in fall time of an expotential pulse when processed through a logarithmic amplifier as shown in Figure 7.

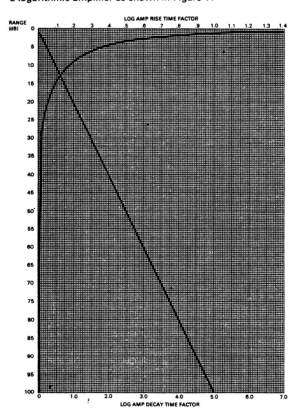

Figure 7

Courtesy of Kevin Redmond, LEL Division of Varian, Copiaque, New York

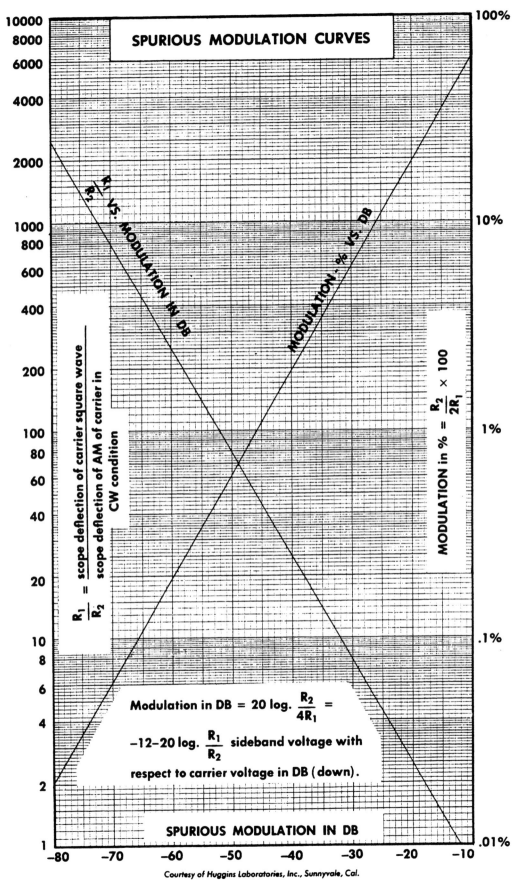

SPURIOUS MODULATION CURVES

$\frac{R_1}{R_2}$ VS. MODULATION IN DB

MODULATION - % VS. DB

$\frac{R_1}{R_2} = \dfrac{\text{scope deflection of carrier square wave}}{\text{scope deflection of AM of carrier in CW condition}}$

$\text{MODULATION in } \% = \dfrac{R_2}{2R_1} \times 100$

Modulation in DB $= 20 \log. \dfrac{R_2}{4R_1} =$

$-12-20 \log. \dfrac{R_1}{R_2}$ sideband voltage with

respect to carrier voltage in DB (down).

SPURIOUS MODULATION IN DB

Courtesy of Huggins Laboratories, Inc., Sunnyvale, Cal.

DIELECTRIC MATERIALS CHART
PROPERTIES AT MICROWAVE FREQUENCIES

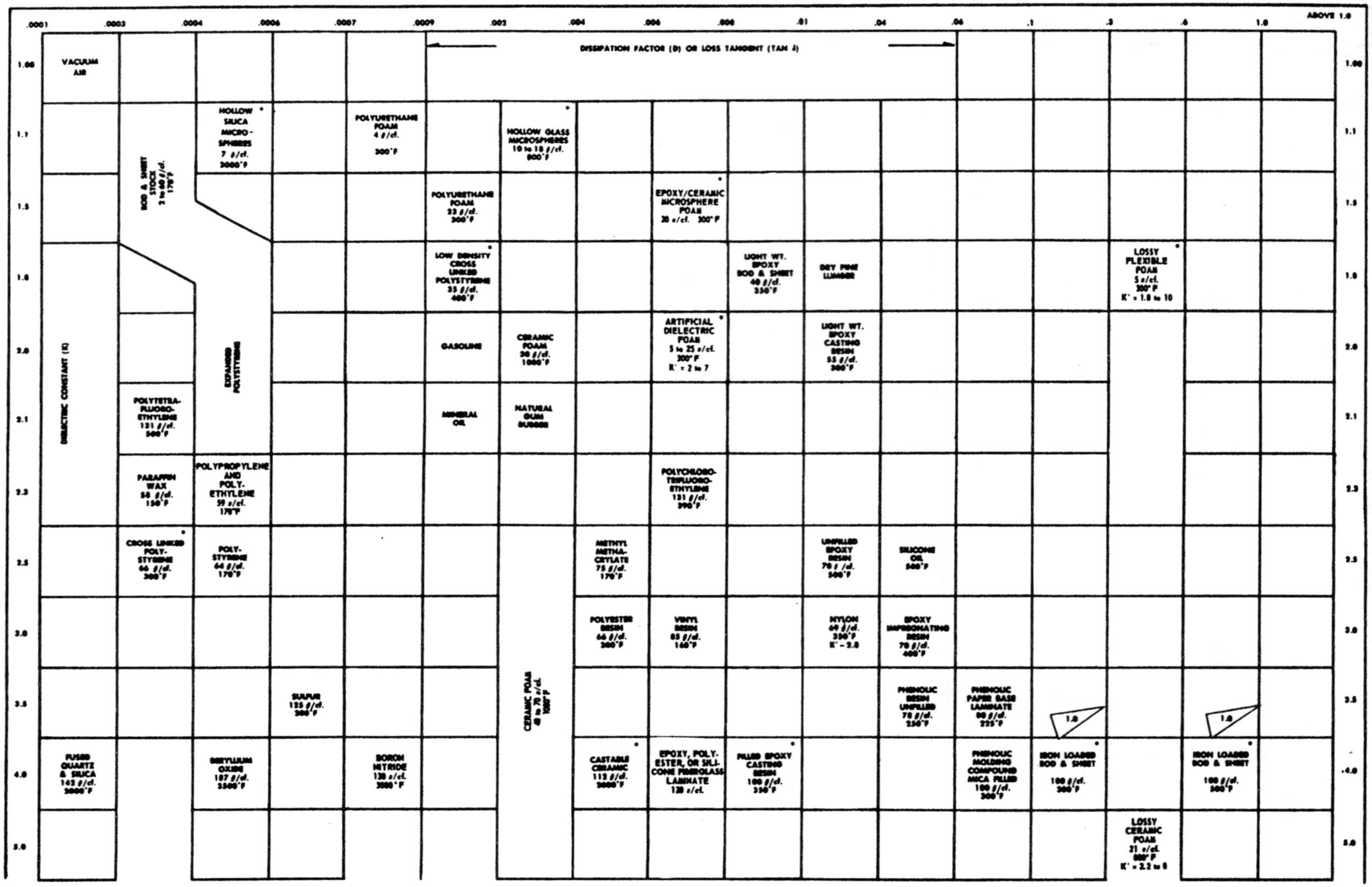

DISSIPATION FACTOR (P) OR LOSS TANGENT (TAN δ)

DIELECTRIC CONSTANT (K)

184

Data for chart provided courtesy Emerson & Cuming Inc., Canton, Mass.

Microwave Engineers'

DEFINITIONS

A non-magnetic dielectric material is defined by real and imaginary components of the Complex Permittivity.

Complex Permittivity

$$\epsilon^* = \epsilon' - j\epsilon'' \text{ (farads/meter)}$$

Normalization of ϵ^* with respect to the Dielectric Permittivity of Free Space.

$$\epsilon_0 = 10^{-9}/36\pi \text{ (farads/meter)}$$

Complex Relative Permittivity gives the

$$\frac{\epsilon^*}{\epsilon_0} = k^* = k' - jk''$$

where

DIELECTRIC CONSTANT k' is the relative permittivity or relative dielectric constant. It is the quantity generally referred to as Dielectric Constant or the Dielectric. It is plotted vertically on this chart and is a dimensionless quantity since it is relative to free space.

LOSS FACTOR k'' is the relative loss factor. It is usually given as Loss Factor. It should not be confused with Dissipation Factor which is the quantity usually used to characterize a material. Plotted horizontally on this chart, it is dimensionless and defined on a relative basis.

DISSIPATION FACTOR Dissipation Factor (D), Loss Tangent and tan δ are identical.

LOSS TANGENT

$$D = \tan \delta = \frac{k''}{k'}$$

tan δ

Note that this Loss Factor is the product of the Dielectric Constant and Dissipation Factor.

POWER FACTOR

$$PF = \cos \theta = \frac{\tan \delta}{\sqrt{1 + \tan^2 \delta}} = \tan \delta \text{ approx.}$$

Power Factor is approximately equal to Dissipation Factor for values below 0.1.

QUALITY FACTOR

$$Q = \frac{1}{\tan \delta}$$

DIELECTRIC CONDUCTIVITY

$$\epsilon_c = 5.5 \times 10^{-11} f k'' \text{ (mho/meter)}$$

where f = frequency. When applied to a dielectric material, it is comparable to the "Q" of circuit theory.

ATTENUATION

$$A = 9.1 \times 10^{-8} f \tan \delta \sqrt{k'} \text{ (db/meter)}$$
approx. This applies to the TEM mode.

WAVE IMPEDANCE

$$Z = 377 \sqrt{1/k'} (1 + j \frac{\tan \delta}{2}) \text{ (ohm)}$$
approx.

INDEX OF REFRACTION

$$n = \sqrt{k'}$$

CHART EXPLANATIONS

Properties are average values for:

Microwave Frequency Range: 10^8 to 3×10^{10} Hz.
(Free Space Wavelength, from 30 cm to 1 cm)
and at
Room Temperature, 77°F (25°C)

Note: Properties of all materials vary with both frequency and temperature. Consult appropriate Technical Bulletins or other source information.

Weight of each material is given in pounds per cubic foot (#/cf). To convert to density in grams per cubic centimeter (g/cc) divide by 62.4.

Safe working limit for each material is given in degrees Fahrenheit (°F). To convert to degrees Centigrade (°C) subtract 32 and multiply result by 5/9.

High Temperature indicates a magnetic material. Permeability at microwave frequency is on the flag. Dissipation factor indicated by position on chart is the sum of the dielectric and magnetic dissipation factors.

Data for chart provided courtesy Emerson & Cuming Inc., Canton, Mass.

SKIN DEPTH

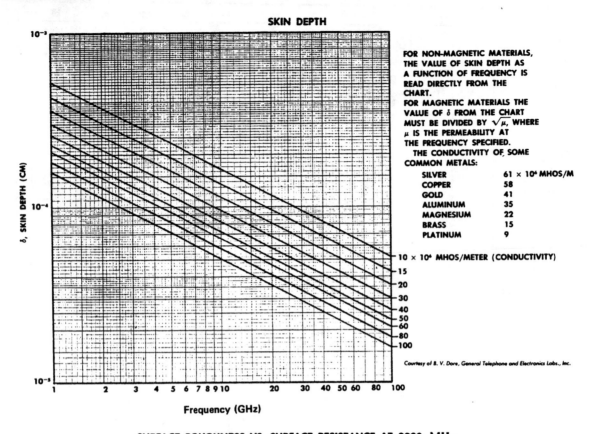

FOR NON-MAGNETIC MATERIALS, THE VALUE OF SKIN DEPTH AS A FUNCTION OF FREQUENCY IS READ DIRECTLY FROM THE CHART.

FOR MAGNETIC MATERIALS THE VALUE OF δ FROM THE CHART MUST BE DIVIDED BY $\sqrt{\mu}$, WHERE μ IS THE PERMEABILITY AT THE FREQUENCY SPECIFIED.

THE CONDUCTIVITY OF SOME COMMON METALS:

SILVER	61×10^4 MHOS/M
COPPER	58
GOLD	41
ALUMINUM	35
MAGNESIUM	22
BRASS	15
PLATINUM	9

Courtesy of B. V. Dore, General Telephone and Electronics Labs., Inc.

Frequency (GHz)

SURFACE ROUGHNESS VS. SURFACE RESISTANCE AT 3000 MHz

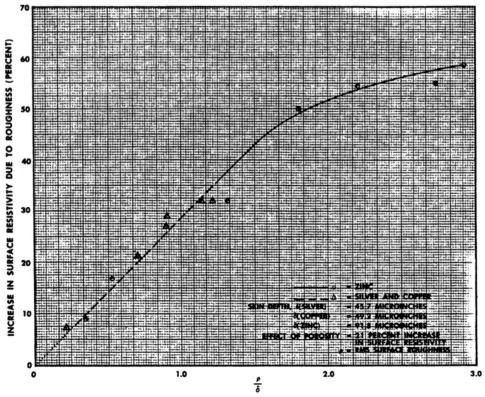

R. Lending, Reprinted from Vol. 11 Proceedings of National Electronics Conference.

Microwave Engineers'

FOUR-PLACE TABLE
DECIBELS RETURN LOSS TO MAGNITUDE OF
VOLTAGE REFLECTION COEFFICIENT

This table gives to four places the magnitude of the voltage reflection coefficient $|\Gamma|$ corresponaing to the return loss $L_R = 20 \log_{10} \dfrac{1}{|\Gamma|}$ for discrete values spaced 0.01 decibel within the range $0 - 20$ decibels.

If L_R is greater than 20 decibels the following procedure will permit determination of $|\Gamma|$.

1. Subtract 20n decibels from L_R, choosing the integer n of such a value so that the result is within the range of the table.

2. Opposite $(L_R - 20n)$ in the table read $|\Gamma_T|$.

3. Shift the decimal place of $|\Gamma_T|$ to the left n times to obtain $|\Gamma|$ corresponding to L_R.
Example: $L_R = 62$ decibels

Choose n = 3, so that
$L_R - 20n = 2$ decibels

Opposite 2 decibels in the table, find
$|\Gamma_T| = 0.7943$

Shifting the decimal point to the
left 3 places,
$|\Gamma| = 0.0007943$.

The table is accurate to the figures shown, being abstracted from seven-place tables.[1]   Other uses for the table may occur whenever the same mathematical relationship is encountered.

Reference

1. R. W. Beatty & W. J. Anson, "Table of Magnitude of Reflection Coefficient Versus Return Loss," NBS Boulder Laboratories Technical Note, No. 72, Sept. 19, 1960. (For sale by Office of Technical Services, U. S. Dept. of Commerce, Washington 25, D.C., Price $1.25.)

187

Courtesy of R. W. Beatty, National Bureau of Standards, Boulder, Colo.

DECIBELS RETURN LOSS TO MAGNITUDE OF VOLTAGE REFLECTION COEFFICIENT

dB	0	1	2	3	4	5	6	7	8	9	dB
0.0	1.0000	.9988	.9977	.9966	.9954	.9943	.9931	.9920	.9908	.9897	0.0
0.1	.9886	.9874	.9863	.9851	.9840	.9829	.9817	.9806	.9795	.9784	0.1
0.2	.9772	.9761	.9750	.9739	.9727	.9716	.9705	.9694	.9683	.9672	0.2
0.3	.9661	.9649	.9638	.9627	.9616	.9605	.9594	.9583	.9572	.9561	0.3
0.4	.9550	.9539	.9528	.9517	.9506	.9495	.9484	.9473	.9462	.9451	0.4
0.5	.9441	.9430	.9419	.9408	.9397	.9386	.9376	.9365	.9354	.9343	0.5
0.6	.9333	.9322	.9311	.9300	.9290	.9279	.9268	.9258	.9247	.9236	0.6
0.7	.9226	.9215	.9204	.9194	.9183	.9173	.9162	.9152	.9141	.9131	0.7
0.8	.9120	.9110	.9099	.9089	.9078	.9068	.9057	.9047	.9036	.9026	0.8
0.9	.9016	.9005	.8995	.8985	.8974	.8964	.8954	.8943	.8933	.8923	0.9
1.0	.8913	.8902	.8892	.8882	.8872	.8861	.8851	.8841	.8831	.8821	1.0
1.1	.8810	.8800	.8790	.8780	.8770	.8760	.8750	.8740	.8730	.8720	1.1
1.2	.8710	.8700	.8690	.8680	.8670	.8660	.8650	.8640	.8630	.8620	1.2
1.3	.8610	.8600	.8590	.8580	.8570	.8561	.8551	.8541	.8531	.8521	1.3
1.4	.8511	.8502	.8492	.8482	.8472	.8463	.8453	.8443	.8433	.8424	1.4
1.5	.8414	.8404	.8395	.8385	.8375	.8366	.8356	.8346	.8337	.8327	1.5
1.6	.8318	.8308	.8299	.8289	.8279	.8270	.8260	.8251	.8241	.8232	1.6
1.7	.8222	.8213	.8204	.8194	.8185	.8175	.8166	.8156	.8147	.8138	1.7
1.8	.8128	.8119	.8110	.8100	.8091	.8082	.8072	.8063	.8054	.8045	1.8
1.9	.8035	.8026	.8017	.8008	.7998	.7989	.7980	.7971	.7962	.7952	1.9
2.0	.7943	.7934	.7925	.7916	.7907	.7898	.7889	.7880	.7870	.7861	2.0
2.1	.7852	.7843	.7834	.7825	.7816	.7807	.7798	.7789	.7780	.7771	2.1
2.2	.7762	.7754	.7745	.7736	.7727	.7718	.7709	.7700	.7691	.7682	2.2
2.3	.7674	.7665	.7656	.7647	.7638	.7630	.7621	.7612	.7603	.7595	2.3
2.4	.7506	.7577	.7568	.7560	.7551	.7542	.7534	.7525	.7516	.7508	2.4
2.5	.7499	.7490	.7482	.7473	.7464	.7456	.7447	.7439	.7430	.7422	2.5
2.6	.7413	.7405	.7396	.7388	.7379	.7371	.7362	.7354	.7345	.7337	2.6
2.7	.7328	.7320	.7311	.7303	.7295	.7286	.7278	.7269	.7261	.7253	2.7
2.8	.7244	.7236	.7228	.7219	.7211	.7203	.7194	.7186	.7178	.7170	2.8
2.9	.7161	.7153	.7145	.7137	.7129	.7120	.7112	.7104	.7096	.7088	2.9
3.0	.7079	.7071	.7063	.7055	.7047	.7039	.7031	.7023	.7015	.7006	3.0
3.1	.6998	.6990	.6982	.6974	.6966	.6958	.6950	.6942	.6934	.6926	3.1
3.2	.6918	.6910	.6902	.6894	.6887	.6879	.6871	.6863	.6855	.6847	3.2
3.3	.6839	.6831	.6823	.6816	.6808	.6800	.6792	.6784	.6776	.6769	3.3
3.4	.6761	.6753	.6745	.6738	.6730	.6722	.6714	.6707	.6699	.6691	3.4
3.5	.6683	.6676	.6668	.6660	.6653	.6645	.6637	.6630	.6622	.6615	3.5
3.6	.6607	.6599	.6592	.6584	.6577	.6569	.6561	.6554	.6546	.6539	3.6
3.7	.6531	.6524	.6516	.6509	.6501	.6494	.6486	.6479	.6471	.6464	3.7
3.8	.6457	.6449	.6442	.6434	.6427	.6419	.6412	.6405	.6397	.6390	3.8
3.9	.6383	.6375	.6368	.6361	.6353	.6346	.6339	.6331	.6324	.6317	3.9
4.0	.6310	.6302	.6295	.6288	.6281	.6273	.6266	.6259	.6252	.6245	4.0
4.1	.6237	.6230	.6223	.6216	.6209	.6202	.6194	.6187	.6180	.6173	4.1
4.2	.6166	.6159	.6152	.6145	.6138	.6131	.6124	.6116	.6109	.6102	4.2
4.3	.6095	.6088	.6081	.6074	.6067	.6060	.6053	.6046	.6039	.6033	4.3
4.4	.6026	.6019	.6012	.6005	.5998	.5991	.5984	.5977	.5970	.5963	4.4
4.5	.5957	.5950	.5943	.5936	.5929	.5922	.5916	.5909	.5902	.5895	4.5
4.6	.5888	.5882	.5875	.5868	.5861	.5855	.5848	.5841	.5834	.5828	4.6
4.7	.5821	.5814	.5808	.5801	.5794	.5788	.5781	.5774	.5768	.5761	4.7
4.8	.5754	.5748	.5741	.5735	.5728	.5721	.5715	.5708	.5702	.5695	4.8
4.9	.5689	.5682	.5675	.5669	.5662	.5656	.5649	.5643	.5636	.5630	4.9
dB	0	1	2	3	4	5	6	7	8	9	dB

Courtesy of R. W. Beatty, National Bureau of Standards, Boulder, Colo.

Microwave Engineers'

DECIBELS RETURN LOSS TO MAGNITUDE OF
VOLTAGE REFLECTION COEFFICIENT

dB	0	1	2	3	4	5	6	7	8	9	dB
5.0	.5623	.5617	.5610	.5604	.5598	.5591	.5585	.5578	.5572	.5565	5.0
5.1	.5559	.5553	.5546	.5540	.5534	.5527	.5521	.5514	.5508	.5502	5.1
5.2	.5495	.5489	.5483	.5476	.5470	.5464	.5458	.5451	.5445	.5439	5.2
5.3	.5433	.5426	.5420	.5414	.5408	.5401	.5395	.5389	.5383	.5377	5.3
5.4	.5370	.5364	.5358	.5352	.5346	.5339	.5333	.5327	.5321	.5315	5.4
5.5	.5309	.5303	.5297	.5291	.5284	.5278	.5272	.5266	.5260	.5254	5.5
5.6	.5248	.5242	.5236	.5230	.5224	.5218	.5212	.5206	.5200	.5194	5.6
5.7	.5188	.5182	.5176	.5170	.5164	.5158	.5152	.5146	.5140	.5135	5.7
5.8	.5129	.5123	.5117	.5111	.5105	.5099	.5093	.5087	.5082	.5076	5.8
5.9	.5070	.5064	.5058	.5052	.5047	.5041	.5035	.5029	.5023	.5018	5.9
6.0	.5012	.5006	.5000	.4995	.4989	.4983	.4977	.4972	.4966	.4960	6.0
6.1	.4955	.4949	.4943	.4937	.4932	.4926	.4920	.4915	.4909	.4903	6.1
6.2	.4898	.4892	.4887	.4881	.4875	.4870	.4864	.4858	.4853	.4847	6.2
6.3	.4842	.4836	.4831	.4825	.4819	.4814	.4808	.4803	.4797	.4792	6.3
6.4	.4786	.4781	.4775	.4770	.4764	.4759	.4753	.4748	.4742	.4737	6.4
6.5	.4732	.4726	.4721	.4715	.4710	.4704	.4699	.4694	.4688	.4683	6.5
6.6	.4677	.4672	.4667	.4661	.4656	.4651	.4645	.4640	.4634	.4629	6.6
6.7	.4624	.4618	.4613	.4608	.4603	.4597	.4592	.4587	.4581	.4576	6.7
6.8	.4571	.4566	.4560	.4555	.4550	.4545	.4539	.4534	.4529	.4524	6.8
6.9	.4519	.4513	.4508	.4503	.4498	.4493	.4487	.4482	.4477	.4472	6.9
7.0	.4467	.4462	.4457	.4451	.4446	.4441	.4436	.4431	.4426	.4421	7.0
7.1	.4416	.4411	.4406	.4400	.4395	.4390	.4385	.4380	.4375	.4370	7.1
7.2	.4365	.4360	.4355	.4350	.4345	.4340	.4335	.4330	.4325	.4320	7.2
7.3	.4315	.4310	.4305	.4300	.4295	.4290	.4285	.4281	.4276	.4271	7.3
7.4	.4266	.4261	.4256	.4251	.4246	.4241	.4236	.4232	.4227	.4222	7.4
7.5	.4217	.4212	.4207	.4202	.4198	.4193	.4188	.4183	.4178	.4173	7.5
7.6	.4169	.4164	.4159	.4154	.4150	.4145	.4140	.4135	.4130	.4126	7.6
7.7	.4121	.4116	.4111	.4107	.4102	.4097	.4093	.4088	.4083	.4078	7.7
7.8	.4074	.4069	.4064	.4060	.4055	.4050	.4046	.4041	.4036	.4032	7.8
7.9	.4027	.4023	.4018	.4013	.4009	.4004	.3999	.3995	.3990	.3986	7.9
8.0	.3981	.3976	.3972	.3967	.3963	.3958	.3954	.3949	.3945	.3940	8.0
8.1	.3936	.3931	.3926	.3922	.3917	.3913	.3908	.3904	.3899	.3895	8.1
8.2	.3890	.3886	.3882	.3877	.3873	.3868	.3864	.3859	.3855	.3850	8.2
8.3	.3846	.3841	.3837	.3833	.3828	.3824	.3819	.3815	.3811	.3806	8.3
8.4	.3802	.3798	.3793	.3789	.3784	.3780	.3776	.3771	.3767	.3763	8.4
8.5	.3758	.3754	.3750	.3745	.3741	.3737	.3733	.3728	.3724	.3720	8.5
8.6	.3715	.3711	.3707	.3703	.3698	.3694	.3690	.3686	.3681	.3677	8.6
8.7	.3673	.3669	.3664	.3660	.3656	.3652	.3648	.3643	.3639	.3635	8.7
8.8	.3631	.3627	.3622	.3618	.3614	.3610	.3606	.3602	.3597	.3593	8.8
8.9	.3589	.3585	.3581	.3577	.3573	.3569	.3565	.3560	.3556	.3552	8.9
9.0	.3548	.3544	.3540	.3536	.3532	.3528	.3524	.3520	.3516	.3512	9.0
9.1	.3508	.3503	.3499	.3495	.3491	.3487	.3483	.3479	.3475	.3471	9.1
9.2	.3467	.3463	.3459	.3455	.3451	.3447	.3443	.3440	.3436	.3432	9.2
9.3	.3428	.3424	.3420	.3416	.3412	.3408	.3404	.3400	.3396	.3392	9.3
9.4	.3388	.3385	.3381	.3377	.3373	.3369	.3365	.3361	.3357	.3354	9.4
9.5	.3350	.3346	.3342	.3338	.3334	.3330	.3327	.3323	.3319	.3315	9.5
9.6	.3311	.3308	.3304	.3300	.3296	.3292	.3289	.3285	.3281	.3277	9.6
9.7	.3273	.3270	.3266	.3262	.3258	.3255	.3251	.3247	.3243	.3240	9.7
9.8	.3236	.3232	.3228	.3225	.3221	.3217	.3214	.3210	.3206	.3203	9.8
9.9	.3199	.3195	.3192	.3188	.3184	.3181	.3177	.3173	.3170	.3166	9.9
dB	0	1	2	3	4	5	6	7	8	9	dB

189

Courtesy of R. W. Beatty, National Bureau of Standards, Boulder, Colo.

DECIBELS RETURN LOSS TO MAGNITUDE OF VOLTAGE REFLECTION COEFFICIENT

dB	0	1	2	3	4	5	6	7	8	9	dB
10.0	.3162	.3159	.3155	.3151	.3148	.3144	.3141	.3137	.3133	.3130	10.0
10.1	.3126	.3122	.3119	.3115	.3112	.3108	.3105	.3101	.3097	.3094	10.1
10.2	.3090	.3087	.3083	.3080	.3076	.3073	.3069	.3065	.3062	.3058	10.2
10.3	.3055	.3051	.3048	.3044	.3041	.3037	.3034	.3030	.3027	.3023	10.3
10.4	.3020	.3016	.3013	.3010	.3006	.3003	.2999	.2996	.2992	.2989	10.4
10.5	.2985	.2982	.2979	.2975	.2972	.2968	.2965	.2961	.2958	.2955	10.5
10.6	.2951	.2948	.2944	.2941	.2938	.2934	.2931	.2928	.2924	.2921	10.6
10.7	.2917	.2914	.2911	.2907	.2904	.2901	.2897	.2894	.2891	.2887	10.7
10.8	.2884	.2881	.2877	.2874	.2871	.2867	.2864	.2861	.2858	.2854	10.8
10.9	.2851	.2848	.2844	.2841	.2838	.2835	.2831	.2828	.2825	.2822	10.9
11.0	.2818	.2815	.2812	.2809	.2805	.2802	.2799	.2796	.2793	.2789	11.0
11.1	.2786	.2783	.2780	.2777	.2773	.2770	.2767	.2764	.2761	.2757	11.1
11.2	.2754	.2751	.2748	.2745	.2742	.2738	.2735	.2732	.2729	.2726	11.2
11.3	.2723	.2720	.2716	.2713	.2710	.2707	.2704	.2701	.2698	.2695	11.3
11.4	.2692	.2688	.2685	.2682	.2679	.2676	.2673	.2670	.2667	.2664	11.4
11.5	.2661	.2658	.2655	.2652	.2649	.2645	.2642	.2639	.2636	.2633	11.5
11.6	.2630	.2627	.2624	.2621	.2618	.2615	.2612	.2609	.2606	.2603	11.6
11.7	.2600	.2597	.2594	.2591	.2588	.2585	.2582	.2579	.2576	.2573	11.7
11.8	.2570	.2567	.2564	.2562	.2559	.2556	.2553	.2550	.2547	.2544	11.8
11.9	.2541	.2538	.2535	.2532	.2529	.2526	.2523	.2521	.2518	.2515	11.9
12.0	.2512	.2509	.2506	.2503	.2500	.2497	.2495	.2492	.2489	.2486	12.0
12.1	.2483	.2480	.2477	.2475	.2472	.2469	.2466	.2463	.2460	.2458	12.1
12.2	.2455	.2452	.2449	.2446	.2443	.2441	.2438	.2435	.2432	.2429	12.2
12.3	.2427	.2424	.2421	.2418	.2415	.2413	.2410	.2407	.2404	.2402	12.3
12.4	.2399	.2396	.2393	.2391	.2388	.2385	.2382	.2380	.2377	.2374	12.4
12.5	.2371	.2369	.2366	.2363	.2360	.2358	.2355	.2352	.2350	.2347	12.5
12.6	.2344	.2342	.2339	.2336	.2333	.2331	.2328	.2325	.2323	.2320	12.6
12.7	.2317	.2315	.2312	.2309	.2307	.2304	.2301	.2299	.2296	.2294	12.7
12.8	.2291	.2288	.2286	.2283	.2280	.2278	.2275	.2272	.2270	.2267	12.8
12.9	.2265	.2262	.2259	.2257	.2254	.2252	.2249	.2246	.2244	.2241	12.9
13.0	.2239	.2236	.2234	.2231	.2228	.2226	.2223	.2221	.2218	.2216	13.0
13.1	.2213	.2211	.2208	.2205	.2203	.2200	.2198	.2195	.2193	.2190	13.1
13.2	.2188	.2185	.2183	.2180	.2178	.2175	.2173	.2170	.2168	.2165	13.2
13.3	.2163	.2160	.2158	.2155	.2153	.2150	.2148	.2145	.2143	.2140	13.3
13.4	.2138	.2136	.2133	.2131	.2128	.2126	.2123	.2121	.2118	.2116	13.4
13.5	.2113	.2111	.2109	.2106	.2104	.2101	.2099	.2097	.2094	.2092	13.5
13.6	.2089	.2087	.2084	.2082	.2080	.2077	.2075	.2073	.2070	.2068	13.6
13.7	.2065	.2063	.2061	.2058	.2056	.2054	.2051	.2049	.2046	.2044	13.7
13.8	.2042	.2039	.2037	.2035	.2032	.2030	.2028	.2025	.2023	.2021	13.8
13.9	.2018	.2016	.2014	.2011	.2009	.2007	.2004	.2002	.2000	.1998	13.9
14.0	.1995	.1993	.1991	.1988	.1986	.1984	.1982	.1979	.1977	.1975	14.0
14.1	.1972	.1970	.1968	.1966	.1963	.1961	.1959	.1957	.1954	.1952	14.1
14.2	.1950	.1948	.1945	.1943	.1941	.1939	.1936	.1934	.1932	.1930	14.2
14.3	.1928	.1925	.1923	.1921	.1919	.1916	.1914	.1912	.1910	.1908	14.3
14.4	.1905	.1903	.1901	.1899	.1897	.1895	.1892	.1890	.1888	.1886	14.4
14.5	.1884	.1881	.1879	.1877	.1875	.1873	.1871	.1869	.1866	.1864	14.5
14.6	.1862	.1860	.1858	.1856	.1854	.1851	.1849	.1847	.1845	.1843	14.6
14.7	.1841	.1839	.1837	.1834	.1832	.1830	.1828	.1826	.1824	.1822	14.7
14.8	.1820	.1818	.1816	.1813	.1811	.1809	.1807	.1805	.1803	.1801	14.8
14.9	.1799	.1797	.1795	.1793	.1791	.1789	.1786	.1784	.1782	.1780	14.9
dB	0	1	2	3	4	5	6	7	8	9	dB

Courtesy of R. W. Beatty, National Bureau of Standards, Boulder, Colo.

Microwave Engineers'

DECIBELS RETURN LOSS TO MAGNITUDE OF VOLTAGE REFLECTION COEFFICIENT

dB	0	1	2	3	4	5	6	7	8	9	dB
15.0	.1778	.1776	.1774	.1772	.1770	.1768	.1766	.1764	.1762	.1760	15.0
15.1	.1758	.1756	.1754	.1752	.1750	.1748	.1746	.1744	.1742	.1740	15.1
15.2	.1738	.1736	.1734	.1732	.1730	.1728	.1726	.1724	.1722	.1720	15.2
15.3	.1718	.1716	.1714	.1712	.1710	.1708	.1706	.1704	.1702	.1700	15.3
15.4	.1698	.1696	.1694	.1692	.1690	.1688	.1687	.1685	.1683	.1681	15.4
15.5	.1679	.1677	.1675	.1673	.1671	.1669	.1667	.1665	.1663	.1661	15.5
15.6	.1660	.1658	.1656	.1654	.1652	.1650	.1648	.1646	.1644	.1642	15.6
15.7	.1641	.1639	.1637	.1635	.1633	.1631	.1629	.1627	.1626	.1624	15.7
15.8	.1622	.1620	.1618	.1616	.1614	.1613	.1611	.1609	.1607	.1605	15.8
15.9	.1603	.1601	.1600	.1598	.1596	.1594	.1592	.1590	.1589	.1587	15.9
16.0	.1585	.1583	.1581	.1579	.1578	.1576	.1574	.1572	.1570	.1569	16.0
16.1	.1567	.1565	.1563	.1561	.1560	.1558	.1556	.1554	.1552	.1551	16.1
16.2	.1549	.1547	.1545	.1543	.1542	.1540	.1538	.1536	.1535	.1533	16.2
16.3	.1531	.1529	.1528	.1526	.1524	.1522	.1521	.1519	.1517	.1515	16.3
16.4	.1514	.1512	.1510	.1508	.1507	.1505	.1503	.1501	.1500	.1498	16.4
16.5	.1496	.1495	.1493	.1491	.1489	.1488	.1486	.1484	.1483	.1481	16.5
16.6	.1479	.1477	.1476	.1474	.1472	.1471	.1469	.1467	.1466	.1464	16.6
16.7	.1462	.1460	.1459	.1457	.1455	.1454	.1452	.1450	.1449	.1447	16.7
16.8	.1445	.1444	.1442	.1440	.1439	.1437	.1435	.1434	.1432	.1431	16.8
16.9	.1429	.1427	.1426	.1424	.1422	.1421	.1419	.1417	.1416	.1414	16.9
17.0	.1413	.1411	.1409	.1408	.1406	.1404	.1403	.1401	.1400	.1398	17.0
17.1	.1396	.1395	.1393	.1392	.1390	.1388	.1387	.1385	.1384	.1382	17.1
17.2	.1380	.1379	.1377	.1376	.1374	.1372	.1371	.1369	.1368	.1366	17.2
17.3	.1365	.1363	.1361	.1360	.1358	.1357	.1355	.1354	.1352	.1351	17.3
17.4	.1349	.1347	.1346	.1344	.1343	.1341	.1340	.1338	.1337	.1335	17.4
17.5	.1334	.1332	.1330	.1329	.1327	.1326	.1324	.1323	.1321	.1320	17.5
17.6	.1318	.1317	.1315	.1314	.1312	.1311	.1309	.1308	.1306	.1305	17.6
17.7	.1303	.1302	.1300	.1299	.1297	.1296	.1294	.1293	.1291	.1290	17.7
17.8	.1288	.1287	.1285	.1284	.1282	.1281	.1279	.1278	.1276	.1275	17.8
17.9	.1274	.1272	.1271	.1269	.1268	.1266	.1265	.1263	.1262	.1260	17.9
18.0	.1259	.1257	.1256	.1255	.1253	.1252	.1250	.1249	.1247	.1246	18.0
18.1	.1245	.1243	.1242	.1240	.1239	.1237	.1236	.1235	.1233	.1232	18.1
18.2	.1230	.1229	.1227	.1226	.1225	.1223	.1222	.1220	.1219	.1218	18.2
18.3	.1216	.1215	.1213	.1212	.1211	.1209	.1208	.1206	.1205	.1204	18.3
18.4	.1202	.1201	.1199	.1198	.1197	.1195	.1194	.1193	.1191	.1190	18.4
18.5	.1189	.1187	.1186	.1184	.1183	.1182	.1180	.1179	.1178	.1176	18.5
18.6	.1175	.1174	.1172	.1171	.1170	.1168	.1167	.1165	.1164	.1163	18.6
18.7	.1161	.1160	.1159	.1157	.1156	.1155	.1153	.1152	.1151	.1149	18.7
18.8	.1148	.1147	.1146	.1144	.1143	.1142	.1140	.1139	.1138	.1136	18.8
18.9	.1135	.1134	.1132	.1131	.1130	.1128	.1127	.1126	.1125	.1123	18.9
19.0	.1122	.1121	.1119	.1118	.1117	.1116	.1114	.1113	.1112	.1110	19.0
19.1	.1109	.1108	.1107	.1105	.1104	.1103	.1102	.1100	.1099	.1098	19.1
19.2	.1096	.1095	.1094	.1093	.1091	.1090	.1089	.1088	.1086	.1085	19.2
19.3	.1084	.1083	.1081	.1080	.1079	.1078	.1076	.1075	.1074	.1073	19.3
19.4	.1072	.1070	.1069	.1068	.1067	.1065	.1064	.1063	.1062	.1060	19.4
19.5	.1059	.1058	.1057	.1056	.1054	.1053	.1052	.1051	.1050	.1048	19.5
19.6	.1047	.1046	.1045	.1044	.1042	.1041	.1040	.1039	.1038	.1036	19.6
19.7	.1035	.1034	.1033	.1032	.1030	.1029	.1028	.1027	.1026	.1024	19.7
19.8	.1023	.1022	.1021	.1020	.1019	.1017	.1016	.1015	.1014	.1013	19.8
19.9	.1012	.1010	.1009	.1008	.1007	.1006	.1005	.1003	.1002	.1001	19.9
dB	0	1	2	3	4	5	6	7	8	9	dB

Courtesy of R. W. Beatty, National Bureau of Standards, Boulder, Colo.

191

Table of First 700 Zeros of Bessel Functions— $J_l(x)$ and $J'_l(x)$

In the table given here the first seven hundred roots of Bessel functions $J_l(x) = 0$ and $J'_l(x) = 0$ have been computed and arranged in the order of the magnitude of the arguments corresponding to the roots. In the table l is the order of the Bessel function and m is the serial number of the zero of either $J_l(x)$ or $J'_l(x)$, not counting $x = 0$. In waveguide applications the zeros of $J_l(x)$ correspond to transverse magnetic modes of propagation (TM modes) and those of $J'_l(x)$ to transverse electric modes (TE modes). The designations TM and TE appear in the table for the benefit of those who will use this table in waveguide research and serve as a code designating $J_l(x)$ and $J'_l(x)$ for those who are interested in a more general application of the mathematics.

The roots of the Bessel functions were calculated from the *Tables of the Bessel Functions of the First Kind of Orders, J_0 through J_{51},* computed by the Staff of the Computation Laboratory of Harvard University, published by the Harvard University Press, 1946–1948.

TABLE

No.	Mode*	l-m	Value†		No.	Mode*	l-m	Value†
1	TE	1-1	1.841184		48	TM	1-4	13.323692
2	TM	0-1	2.404826		49	TE	0-4	13.323692
3	TE	2-1	3.054237		50	TM	9-1	13.354300
(4	TM	1-1	3.831706		51	TM	6-2	13.589290
(5	TE	0-1	3.831706		52	TE	12-1	13.878843
6	TE	3-1	4.201189		53	TE	5-3	13.987189
7	TM	2-1	5.135622		54	TE	8-2	14.115519
8	TE	4-1	5.317553		55	TM	4-3	14.372537
9	TE	1-2	5.331443		56	TM	10-1	14.475501
10	TM	0-2	5.520078		57	TE	3-4	14.585848
11	TM	3-1	6.380162		58	TM	2-4	14.795952
12	TE	5-1	6.415616		59	TM	7-2	14.821269
13	TE	2-2	6.706133		60	TE	1-5	14.863589
(14	TM	1-2	7.015587		61	TE	13-1	14.928374
(15	TE	0-2	7.015587		62	TM	0-5	14.930918
16	TE	6-1	7.501266		63	TE	6-3	15.268181
17	TM	4-1	7.588342		64	TE	9-2	15.286738
18	TE	3-2	8.015237		65	TM	11-1	15.589848
19	TM	2-2	8.417244		66	TM	5-3	15.700174
20	TE	1-3	8.536316		67	TE	4-4	15.964107
21	TE	7-1	8.577836		68	TE	14-1	15.975439
22	TM	0-3	8.653728		69	TM	8-2	16.037774
23	TM	5-1	8.771484		70	TM	3-4	16.223466
24	TE	4-2	9.282396		71	TE	2-5	16.347322
25	TE	8-1	9.647422		72	TE	10-2	16.447853
26	TM	3-2	9.761023		73	TM	1-5	16.470630
27	TM	6-1	9.936110		74	TE	0-5	16.470630
28	TE	2-3	9.969468		75	TM	6-3	16.529366
(29	TM	1-3	10.173468		76	TM	12-1	16.698250
(30	TE	0-3	10.173468		77	TE	7-3	17.003820
31	TE	5-2	10.519861		78	TE	15-1	17.020323
32	TE	9-1	10.711434		79	TM	9-2	17.241220
33	TM	4-2	11.064709		80	TE	5-4	17.312842
34	TM	7-1	11.086370		81	TE	11-2	17.600267
35	TE	3-3	11.345924		82	TM	4-4	17.615966
36	TM	2-3	11.619841		83	TE	8-3	17.774012
37	TE	1-4	11.706005		84	TE	3-5	17.788748
38	TE	6-2	11.734936		85	TM	13-1	17.801435
39	TE	10-1	11.770877		86	TM	2-5	17.959819
40	TM	0-4	11.791534		87	TE	1-6	18.015528
41	TM	8-1	12.225092		88	TE	16-1	18.063261
42	TM	5-2	12.338604		89	TM	0-6	18.071064
43	TE	4-3	12.681908		90	TM	7-3	18.287583
44	TE	11-1	12.826491		91	TM	10-2	18.433464
45	TE	7-2	12.932386		92	TE	6-4	18.637443
46	TM	3-3	13.015201		93	TE	9-3	18.745091
47	TE	2-4	13.170371		94	TM	14-1	18.899998

*TM designates a zero of $J_l(x)$; TE designates a zero of $J'_l(x)$; in each case l corresponds to the order of the Bessel function and m is the number of the root.
†5 in last place indicates higher value and 5 indicates lower value in rounding off for fewer decimal places.

Courtesy of Bell Telephone Labs., Inc., Whippany, N. J.

Bessel Functions—Continued

TABLE — Continued

No.	Mode*	l-m	Value†
95	TM	5-4	18.980134
96	TE	9-3	19.004594
97	TE	17-1	19.104458
98	TE	4-5	19.196029
99	TM	3-5	19.409415
100	TM	2-6	19.512913
101	TM	8-3	19.554536
102	TM	1-6	19.615859
(103)	TE	0-6	19.615859
104	TM	11-2	19.615967
105	TE	13-2	19.883224
106	TE	7-4	19.941853
107	TM	15-1	19.994431
108	TE	18-1	20.144079
109	TE	10-3	20.223031
110	TM	6-4	20.320789
111	TE	5-5	20.575512
112	TM	12-2	20.789906
113	TM	9-3	20.807048
114	TM	4-5	20.826933
115	TE	3-6	20.972477
116	TE	14-2	21.015405
117	TE	16-1	21.085147
118	TM	2-6	21.116997
119	TE	1-7	21.164370
120	TE	19-1	21.182267
121	TE	0-7	21.211637
122	TM	8-4	21.229063
123	TE	11-3	21.430854
124	TM	7-4	21.641541
125	TE	6-5	21.931715
126	TM	13-3	21.956244
127	TM	10-3	22.046785
128	TE	15-2	22.142247
129	TM	17-1	22.172495
130	TM	5-5	22.217800
131	TE	20-1	22.219145
132	TE	4-6	22.401032
133	TE	9-4	22.501399
134	TM	3-6	22.582730
135	TM	12-3	22.629300
136	TE	2-7	22.671582
137	TM	1-7	22.760084
(138)	TE	0-7	22.760084
139	TM	8-4	22.945173
140	TE	14-2	23.115778
141	TM	21-1	23.254816
142	TE	25-1	23.256777
143	TM	16-2	23.264269
144	TE	18-1	23.268053
145	TE	13-4	23.275854
146	TM	6-5	23.586084
147	TE	10-4	23.760716
148	TE	5-6	23.803581
149	TE	13-3	23.819374
150	TM	4-6	24.019020
151	TE	3-7	24.144897
152	TM	9-4	24.233885
153	TM	15-2	24.269180
154	TE	2-7	24.270112
155	TE	22-1	24.289385
156	TE	1-8	24.311327
157	TM	19-1	24.338250
158	TM	0-8	24.352472
159	TE	17-2	24.381913
160	TM	12-3	24.494885
161	TE	8-5	24.587197
162	TM	7-5	24.934928
163	TE	14-3	25.001972
164	TM	11-4	25.008519
165	TE	6-6	25.183925
166	TE	23-1	25.322921
167	TM	16-2	25.417019
168	TE	20-1	25.417141
169	TE	5-6	25.430341
170	TE	18-2	25.495508
171	TM	10-4	25.509450
172	TE	4-7	25.589760
173	TM	3-7	25.705104
174	TM	0-8	25.748167
175	TE	9-5	25.826031
176	TE	1-8	25.891177
(177)	TM	7-5	25.903672
(178)	TM	13-4	25.903672
179	TE	10-3	26.177766
180	TE	15-3	26.246048
181	TM	12-4	26.266815
182	TE	5-7	26.355506
183	TE	24-1	26.493648
184	TM	21-1	26.545032
185	TE	7-6	26.559784
186	TE	17-2	26.605533
187	TE	19-2	26.773323
188	TM	11-4	26.820152
189	TE	6-6	26.907369
190	TM	14-3	27.010308
191	TE	3-8	27.182022
192	TM	8-5	27.199088
193	TE	25-1	27.310058
194	TM	2-8	27.347386
195	TM	1-9	27.387204
196	TE	13-4	27.420574
197	TM	6-5	27.457051
198	TE	13-4	27.474340
199	TM	22-1	27.493480
200	TM	2-8	27.567944
201	TM	9-5	27.583749
202	TM	18-2	27.697899
203	TE	20-2	27.712126
204	TE	8-6	27.889270
205	TM	12-4	28.026710
206	TM	15-3	28.102416
207	TE	7-6	28.191189
208	TE	6-7	28.409776
209	TE	26-1	28.418072
210	TE	11-5	28.460857
211	TE	17-3	28.511361
212	TM	5-7	28.626619
213	TM	23-1	28.640185
214	TE	14-4	28.694271
215	TE	4-8	28.767836
216	TE	21-2	28.815590
217	TM	19-2	28.831731
218	TM	10-5	28.887375
219	TM	3-8	28.908351
220	TE	2-9	28.977673
221	TM	1-9	29.044829
(222)	TE	0-9	29.046829
223	TE	9-6	29.218564
224	TM	13-4	29.270631
225	TE	16-3	29.290871
226	TE	27-1	29.448163
227	TM	8-6	29.545660
228	TE	18-3	29.670147
229	TM	24-1	29.710509
230	TE	7-7	29.728978
231	TE	12-6	29.790749
232	TE	0-8	29.906591
233	TE	22-2	29.916147
234	TM	20-2	29.961604
235	TM	6-7	30.033723
236	TE	11-5	30.179061
237	TE	5-8	30.202849
238	TM	4-8	30.371008
239	TE	3-9	30.470269
240	TE	17-3	30.473280
241	TE	28-1	30.477523
242	TM	14-4	30.505951
243	TE	10-6	30.534505
244	TM	2-9	30.569205
245	TE	1-10	30.601923
246	TE	0-10	30.634607
247	TM	25-1	30.779039
248	TE	19-3	30.824148
249	TM	9-6	30.885379
250	TE	13-5	30.987394
251	TE	23-2	31.013998
252	TM	21-2	31.087805
253	TE	16-4	31.111945
254	TE	8-7	31.155327
255	TM	7-7	31.422795
256	TM	12-5	31.459960
257	TE	29-1	31.506195
258	TE	6-8	31.617876
259	TM	18-3	31.650118
260	TM	15-4	31.733414
261	TE	5-8	31.811717
262	TE	11-6	31.838425
263	TM	26-1	31.845888
264	TE	4-9	31.938540
265	TE	20-3	31.973715
266	TM	3-9	32.064853
267	TE	24-2	32.109320
268	TE	2-10	32.127327
(269)	TM	1-10	32.189680
(270)	TE	0-10	32.189680
271	TM	22-2	32.210587
272	TM	10-6	32.211856
273	TE	14-5	32.236970
274	TE	17-4	32.310894
275	TE	30-1	32.505248
276	TM	13-5	32.534220
277	TE	13-5	32.731053
278	TE	8-7	32.795800
279	TM	19-3	32.821803
280	TM	27-1	32.911154
281	TM	16-4	32.953665
282	TE	7-8	33.015179
283	TE	21-3	33.119162
284	TE	12-6	33.131450
285	TE	25-2	33.202272
286	TE	6-8	33.233042
287	TM	23-2	33.330177
288	TE	5-9	33.385444
289	TE	15-5	33.478449
290	TE	18-4	33.503929
291	TM	11-6	33.526364
292	TE	4-9	33.537138
293	TE	31-1	33.561634
294	TE	3-10	33.626949
295	TM	2-10	33.716520
296	TE	1-11	33.746183
297	TE	0-11	33.775821
298	TE	10-7	33.841966
299	TE	28-1	33.974930
300	TM	20-3	33.988703
301	TM	14-5	33.99319
302	TM	9-7	34.15438
303	TE	17-4	34.16727
304	TM	22-3	34.26077
305	TE	26-2	34.29300
306	TE	8-8	34.39663

*TM designates a zero of $J_l(x)$; TE designates a zero of $J'_l(x)$; in each case l corresponds to the order of the Bessel function and m is the number of the root.

†5 in last place indicates higher value and $\underline{5}$ indicates lower value in rounding off for fewer decimal places.

Courtesy of Bell Telephone Labs., Inc., Whippany, N. J.

Bessel Functions—Continued

TABLE—Continued

	Mode*	l-m	Value†
307	TE	13-6	34.41455
308	TM	24-2	34.44678
309	TE	32-1	34.58847
310	TM	7-8	34.63709
311	TE	19-4	34.69148
312	TE	16-5	34.71248
313	TE	6-9	34.81339
314	TM	12-6	34.82999
315	TM	5-9	34.98878
316	TM	29-1	35.03730
317	TE	4-10	35.10392
318	TM	21-3	35.15115
319	TE	11-7	35.1667
320	TM	3-10	35.21867
321	TM	15-5	35.24709
322	TE	2-11	35.27554
(323)	TM	1-11	35.33231
(324)	TE	0-11	35.33231
325	TM	18-4	35.37472
326	TE	27-2	35.38163
327	TE	23-3	35.39878
328	TM	10-7	35.49991
329	TM	25-2	35.56057
330	TE	33-1	35.61475
331	TE	14-6	35.68854
332	TE	9-8	35.76379
333	TM	20-4	35.87394
334	TE	17-5	35.93963
335	TM	8-8	36.02562
336	TM	30-1	36.09834
337	TM	13-6	36.12366
338	TE	7-9	36.22438
339	TM	22-3	36.30943
340	TM	6-9	36.42202
341	TE	28-2	36.46829
342	TE	12-7	36.48055
343	TM	16-5	36.49340
344	TE	24-3	36.53343
345	TE	5-10	36.56078
346	TM	19-4	36.57645
347	TE	34-1	36.64051
348	TM	26-2	36.67173
349	TE	4-10	36.69900
350	TE	3-11	36.78102
351	TM	11-7	36.83357
352	TE	2-11	36.86286
353	TE	1-12	36.88999
354	TM	0-12	36.91710
355	TM	15-6	36.95417
356	TE	21-4	37.05164
357	TE	10-8	37.11800
358	TM	31-1	37.15811
359	TE	18-5	37.16040
360	TM	9-8	37.40010
361	TM	14-6	37.40819
362	TM	23-3	37.46381
363	TM	29-2	37.55307
364	TE	8-9	37.62008
365	TE	25-3	37.66491
366	TE	35-1	37.66577
367	TM	17-5	37.73268
368	TM	20-4	37.77286
369	TE	27-2	37.78040
370	TE	13-7	37.78438
371	TM	7-9	37.83872
372	TE	6-10	37.99964
373	TM	12-7	38.15638
374	TM	5-10	38.15987
375	TE	16-6	38.21206
376	TM	32-1	38.21669
377	TE	22-4	38.22490
378	TE	4-11	38.26532
379	TM	3-11	38.37047
380	TE	19-5	38.37524
381	TM	2-12	38.42266
382	TE	11-8	38.46039
(383)	TM	1-12	38.47477
(384)	TE	0-12	38.47477
385	TM	24-3	38.61452
386	TE	30-2	38.63609
387	TM	15-6	38.68428
388	TE	36-1	38.69055
389	TM	10-8	38.76181
390	TE	26-3	38.79341
391	TE	28-2	38.88671
392	TM	21-4	38.96429
393	TE	18-5	38.96543
394	TE	9-9	39.00190
395	TE	14-7	39.07900
396	TM	8-9	39.24045
397	TM	33-1	39.27413
398	TE	23-4	39.39398
399	TE	7-10	39.42227
400	TE	17-6	39.46277
401	TM	13-7	39.46921
402	TM	20-5	39.58453
403	TM	6-10	39.60324
404	TE	37-1	39.71489
405	TM	31-2	39.71743
406	TE	5-11	39.73064
407	TM	25-3	39.76179
408	TE	12-8	39.79194

*TM designates a zero of $J_l(x)$; TE designates a zero of $J'_l(x)$, in each case l corresponds to the order of the Bessel function and m is the number of the root.
†5 in last place indicates higher value and 5 indicates lower value in rounding off for fewer decimal places.

TABLE—Continued

	Mode*	l-m	Value†
409	TM	4-11	39.85763
410	TE	27-3	39.91909
411	TE	3-12	39.93311
412	TM	16-6	39.95255
413	TM	29-2	39.99080
414	TM	2-12	40.00845
415	TE	1-13	40.03344
416	TM	0-13	40.05843
417	TM	11-8	40.11182
418	TM	22-4	40.15105
419	TM	19-5	40.19210
420	TM	34-1	40.33048
421	TE	15-7	40.36510
422	TE	10-9	40.37107
423	TM	24-4	40.55913
424	TM	9-9	40.62855
425	TM	18-6	40.70680
426	TE	38-1	40.73879
427	TM	14-7	40.77283
428	TE	21-5	40.78864
429	TE	32-2	40.79718
430	TE	8-10	40.83018
431	TM	26-3	40.90580
432	TM	7-10	41.03077
433	TE	28-3	41.04211
434	TM	30-2	41.09278
435	TE	13-8	41.11351
436	TE	6-11	41.17885
437	TM	17-6	41.21357
438	TM	5-11	41.32638
439	TE	23-4	41.33343
440	TM	35-1	41.38580
441	TM	20-5	41.41307
442	TE	4-12	41.42367
443	TM	12-8	41.45109
444	TE	3-12	41.52072
445	TE	2-13	41.56894
(446)	TM	1-13	41.61709
(447)	TE	0-13	41.61709
448	TE	16-7	41.64331
449	TE	25-4	41.72059
450	TE	11-9	41.72863
451	TM	39-1	41.76228
452	TE	33-2	41.87540
453	TE	19-6	41.94459
454	TM	22-5	41.98788
455	TM	10-9	42.00419
456	TE	27-3	42.04674
457	TM	15-7	42.06792
458	TE	29-3	42.16260
459	TM	31-2	42.19275
460	TE	9-10	42.22464
461	TE	14-8	42.42585
462	TM	36-1	42.44014
463	TM	8-10	42.44389
464	TM	18-6	42.46781
465	TM	24-4	42.51168
466	TE	7-11	42.61152
467	TM	21-5	42.62870
468	TM	6-11	42.77848
469	TM	13-8	42.78044
470	TE	40-1	42.78537
471	TE	26-4	42.87855
472	TE	5-12	42.89627
473	TE	17-7	42.91415
474	TE	34-2	42.95218
475	TM	4-12	43.01374
476	TE	12-9	43.07549
477	TE	3-13	43.08363
478	TM	2-13	43.15345
479	TE	20-6	43.17654
480	TE	1-14	43.17663
481	TE	23-5	43.18255
482	TM	28-3	43.18477
483	TM	0-14	43.19979
484	TE	30-3	43.28071
485	TM	32-2	43.29081
486	TM	16-7	43.35507
487	TM	11-9	43.36836
488	TE	5-11	43.49352
489	TE	10-10	43.60677
490	TM	25-4	43.68603
491	TM	19-6	43.71571
492	TE	15-8	43.72963
493	TE	41-1	43.80808
494	TM	22-5	43.83932
495	TE	9-10	43.84380
496	TE	35-2	44.02758
497	TE	8-11	44.03001
498	TE	27-4	44.03321
499	TM	14-8	44.10059
500	TE	18-7	44.17813
501	TM	7-11	44.21541
502	TM	29-3	44.32003
503	TE	6-12	44.35258
504	TE	24-5	44.37290
505	TM	33-2	44.38706
506	TE	31-3	44.39653
507	TM	21-6	44.40300
508	TE	13-9	44.41243

*TM designates a zero of $J_l(x)$; TE designates a zero of $J'_l(x)$, in each case l corresponds to the order of the Bessel function and m is the number of the root.
†5 in last place indicates higher value and 5 indicates lower value in rounding off for fewer decimal places.

Courtesy of Bell Telephone Labs, Inc., Whippany, N. J.

TABLE — Continued

No.	Mode* l-m	Value†	No.	Mode* l-m	Value†
509	TM 5-12	44.48932	558	TM 11-10	46.60813
510	TM 38-1	44.54601	559	TE 33-3	46.62177
511	TE 4-13	44.57762	560	TM 40-1	46.64841
512	TM 17-7	44.63483	561	TE 20-7	46.68717
513	TM 3-13	44.66974	562	TM 16-8	46.71581
514	TE 2-14	44.71455	563	TE 26-5	46.74158
515	TM 12-9	44.72194	564	TE 10-11	46.82896
516	TM 1-14	44.75932	565	TE 23-6	46.84075
(517)	TE 0-14	44.75932	566	TE 44-1	46.87409
518	TE 42-1	44.83043	567	TM 9-11	47.04870
519	TM 26-4	44.85670	568	TE 15-9	47.05946
520	TM 20-6	44.95768	569	TM 19-7	47.17400
521	TE 11-10	44.97753	570	TM 28-4	47.18775
522	TE 16-8	45.02543	571	TE 8-12	47.22176
523	TM 23-5	45.04521	572	TE 38-2	47.24608
524	TE 36-2	45.10166	573	TM 7-12	47.39417
525	TE 28-4	45.18473	574	TM 14-9	47.40035
526	TM 10-10	45.23157	575	TM 22-6	47.42517
527	TM 15-8	45.41219	576	TM 25-5	47.44385
528	TE 9-11	45.43548	577	TE 30-4	47.47899
529	TE 19-7	45.43567	578	TE 6-13	47.52196
530	TM 30-3	45.45267	579	TE 18-8	47.59513
531	TM 34-2	45.48156	580	TM 5-13	47.64940
532	TE 32-3	45.51018	581	TM 36-2	47.66568
533	TE 25-5	45.55917	582	TE 13-10	47.68825
534	TM 39-1	45.59762	583	TM 41-1	47.69840
535	TE 22-6	45.62431	584	TM 32-3	47.71055
536	TM 8-11	45.63844	585	TM 34-3	47.73138
537	TE 14-9	45.74024	586	TE 4-14	47.73367
538	TE 7-12	45.79400	587	TE 3-14	47.81779
539	TE 43-1	45.85243	588	TE 2-15	47.85964
540	TM 18-7	45.90766	589	TE 45-1	47.89542
541	TM 6-12	45.94902	590	TM 1-15	47.90146
542	TM 27-4	46.02388	(591)	TE 0-15	47.90146
543	TE 5-13	46.05857	592	TE 27-5	47.92033
544	TM 13-9	46.06571	593	TE 21-7	47.93298
545	TM 4-13	46.16785	594	TM 12-10	47.97429
546	TE 37-2	46.17447	595	TM 17-8	48.01196
547	TM 21-6	46.19406	596	TE 24-6	48.05260
548	TE 3-14	46.23297	597	TE 11-11	48.21133
549	TM 24-5	46.24664	598	TM 39-2	48.31652
550	TM 2-14	46.29800	599	TM 29-4	48.34846
551	TE 17-8	46.31377	600	TE 16-9	48.37069
552	TE 1-15	46.31960	601	TM 20-7	48.43424
553	TM 29-4	46.33328	602	TM 10-11	48.44715
554	TE 12-10	46.33777	603	TE 31-4	48.62201
555	TE 0-15	46.34119	604	TE 9-12	48.63692
556	TM 35-2	46.57441	605	TM 26-5	48.63706
557	TM 31-3	46.58280	606	TM 23-6	48.65132

*TM designates a zero of $J_l(x)$; TE designates a zero of $J'_l(x)$; in each case l corresponds to the order of the Bessel function and m is the number of the root. †5 in last place indicates higher value and 5 indicates lower value in rounding off for fewer decimal places.

TABLE — Continued

No.	Mode* l-m	Value†	No.	Mode* l-m	Value†
607	TM 15-9	48.72646	654	TM 19-8	50.58367
608	TM 42-1	48.74762	655	TM 31-4	50.66103
609	TM 37-2	48.75542	656	TM 14-10	50.67824
610	TM 8-12	48.82593	657	TE 6-14	50.68782
611	TM 33-3	48.83603	658	TM 5-14	50.80717
612	TE 35-3	48.83910	659	TM 44-1	50.84387
613	TE 19-8	48.86993	660	TE 4-15	50.88616
614	TE 46-1	48.91645	661	TM 33-4	50.90045
615	TE 7-13	48.97107	662	TE 39-2	50.93060
616	TE 14-10	49.02964	663	TE 48-1	50.93760
617	TE 28-5	49.09560	664	TM 22-7	50.93776
618	TM 6-13	49.11577	665	TE 13-11	50.94585
619	TE 22-7	49.17342	666	TM 3-15	50.96503
620	TE 5-14	49.21817	667	TE 2-16	50.97113
621	TE 25-6	49.26009	668	TE 18-9	51.00430
622	TM 18-8	49.30111	669	TM 28-5	51.01228
623	TM 4-14	49.32036	670	TM 1-16	51.04354
624	TM 13-10	49.33078	(671)	TE 0-16	51.04354
625	TE 3-15	49.38130	672	TE 37-3	51.04919
626	TE 40-2	49.38586	673	TM 25-6	51.08055
627	TM 2-15	49.44216	674	TM 35-3	51.08975
628	TE 1-16	49.46239	675	TM 12-11	51.21197
629	TM 0-16	49.48261	676	TE 17-9	51.35527
630	TM 30-4	49.50618	677	TM 21-8	51.40137
631	TE 12-11	49.58340	678	TE 11-12	51.43311
632	TE 17-9	49.67443	679	TE 30-5	51.43637
633	TM 21-7	49.68872	680	TE 42-2	51.52135
634	TE 32-4	49.76246	681	TM 24-7	51.63937
635	TE 43-1	49.79610	682	TM 10-12	51.65325
636	TM 11-11	49.82648	683	TE 27-6	51.66288
637	TM 27-5	49.83465	684	TE 16-10	51.68742
638	TM 38-2	49.84371	685	TM 32-4	51.81316
639	TM 24-6	49.87276	686	TE 9-13	51.83078
640	TE 47-1	49.90146	687	TM 20-8	51.86002
641	TE 36-3	49.94501	688	TE 45-1	51.89095
642	TM 34-3	49.95933	689	TE 49-1	51.97776
643	TE 10-12	50.04043	690	TM 40-2	52.00769
644	TM 16-9	50.04461	691	TM 15-10	52.01615
645	TM 20-8	50.13856	692	TM 34-4	52.01724
646	TM 9-12	50.24533	693	TE 7-14	52.03608
647	TE 29-5	50.26756	694	TE 38-3	52.14375
648	TE 15-10	50.36251	695	TM 23-7	52.15171
649	TE 8-13	50.40702	696	TE 29-5	52.18166
650	TE 23-7	50.40880	697	TM 36-3	52.19465
651	TM 41-2	50.45412	698	TM 41-2	52.19978
652	TE 26-6	50.46345	699	TE 19-9	52.26121
653	TE 7-13	50.56818	700	TE 6-14	52.27945

*TM designates a zero of $J_l(x)$; TE designates a zero of $J'_l(x)$; in each case l corresponds to the order of the Bessel function and m is the number of the root. †5 in last place indicates higher value and 5 indicates lower value in rounding off for fewer decimal places.

Courtesy of Bell Telephone Labs. Inc., Whippany, N. J.

DECIBELS vs. VOLTAGE AND POWER

The Decibel Chart below indicates DB for any ratio of voltage or power up to 100 DB. For voltage ratios greater than 10 (or power ratios greater than 100) the ratio can be broken down into two products, the DB found for each separately, the two results then added. For example: To convert a voltage ratio of 200:1 to DB: 200:1 VR equals the product of 100:1 and 2:1. 100:1 equals 40 DB; 2:1 equals 6 DB. Therefore, 200:1 VR equals 40 DB + 6 DB or 46 DB.

Voltage Ratio	Power Ratio	−db+	Voltage Ratio	Power Ratio	Voltage Ratio	Power Ratio	−db+	Voltage Ratio	Power Ratio	Voltage Ratio	Power Ratio	−db+	Voltage Ratio	Power Ratio
1.0000	1.0000	0	1.000	1.000	.4467	.1995	7.0	2.239	5.012	.1995	.03981	14.0	5.012	25.12
.9886	.9772	.1	1.012	1.023	.4416	.1950	7.1	2.265	5.129	.1972	.03890	14.1	5.070	25.70
.9772	.9550	.2	1.023	1.047	.4365	.1905	7.2	2.291	5.248	.1950	.03802	14.2	5.129	26.30
.9661	.9333	.3	1.035	1.072	.4315	.1862	7.3	2.317	5.370	.1928	.03715	14.3	5.188	26.92
.9550	.9120	.4	1.047	1.096	.4266	.1820	7.4	2.344	5.495	.1905	.03631	14.4	5.248	27.54
.9441	.8913	.5	1.059	1.122	.4217	.1778	7.5	2.371	5.623	.1884	.03548	14.5	5.309	28.18
.9333	.8710	.6	1.072	1.148	.4169	.1738	7.6	2.399	5.754	.1862	.03467	14.6	5.370	28.84
.9226	.8511	.7	1.084	1.175	.4121	.1698	7.7	2.427	5.888	.1841	.03388	14.7	5.433	29.51
.9120	.8318	.8	1.096	1.202	.4074	.1660	7.8	2.455	6.026	.1820	.03311	14.8	5.495	30.20
.9016	.8128	.9	1.109	1.230	.4027	.1622	7.9	2.483	6.166	.1799	.03236	14.9	5.559	30.90
.8913	.7943	1.0	1.122	1.259	.3981	.1585	8.0	2.512	6.310	.1778	.03162	15.0	5.623	31.62
.8810	.7762	1.1	1.135	1.288	.3936	.1549	8.1	2.541	6.457	.1758	.03090	15.1	5.689	32.36
.8710	.7586	1.2	1.148	1.318	.3890	.1514	8.2	2.570	6.607	.1738	.03020	15.2	5.754	33.11
.8610	.7413	1.3	1.161	1.349	.3846	.1479	8.3	2.600	6.761	.1718	.02951	15.3	5.821	33.88
.8511	.7244	1.4	1.175	1.380	.3802	.1445	8.4	2.630	6.918	.1698	.02884	15.4	5.888	34.67
.8414	.7079	1.5	1.189	1.413	.3758	.1413	8.5	2.661	7.079	.1679	.02818	15.5	5.957	35.48
.8318	.6918	1.6	1.202	1.445	.3715	.1380	8.6	2.692	7.244	.1660	.02754	15.6	6.026	36.31
.8222	.6761	1.7	1.216	1.479	.3673	.1349	8.7	2.723	7.413	.1641	.02692	15.7	6.095	37.15
.8128	.6607	1.8	1.230	1.514	.3631	.1318	8.8	2.754	7.586	.1622	.02630	15.8	6.166	38.02
.8035	.6457	1.9	1.245	1.549	.3589	.1288	8.9	2.786	7.762	.1603	.02570	15.9	6.237	38.90
.7943	.6310	2.0	1.259	1.585	.3548	.1259	9.0	2.818	7.943	.1585	.02512	16.0	6.310	39.81
.7852	.6166	2.1	1.274	1.622	.3508	.1230	9.1	2.851	8.128	.1567	.02455	16.1	6.383	40.74
.7762	.6026	2.2	1.288	1.660	.3467	.1202	9.2	2.884	8.318	.1549	.02399	16.2	6.457	41.69
.7674	.5888	2.3	1.303	1.698	.3428	.1175	9.3	2.917	8.511	.1531	.02344	16.3	6.531	42.66
.7586	.5754	2.4	1.318	1.738	.3388	.1148	9.4	2.951	8.710	.1514	.02291	16.4	6.607	43.65
.7499	.5623	2.5	1.334	1.778	.3350	.1122	9.5	2.985	8.913	.1496	.02239	16.5	6.683	44.67
.7413	.5495	2.6	1.349	1.820	.3311	.1096	9.6	3.020	9.120	.1479	.02188	16.6	6.761	45.71
.7328	.5370	2.7	1.365	1.862	.3273	.1072	9.7	3.055	9.333	.1462	.02138	16.7	6.839	46.77
.7244	.5248	2.8	1.380	1.905	.3236	.1047	9.8	3.090	9.550	.1445	.02089	16.8	6.918	47.86
.7161	.5129	2.9	1.396	1.950	.3199	.1023	9.9	3.126	9.772	.1429	.02042	16.9	6.998	48.98
.7079	.5012	3.0	1.413	1.995	.3162	.1000	10.0	3.162	10.000	.1413	.01995	17.0	7.079	50.12
.6998	.4898	3.1	1.429	2.042	.3126	.09772	10.1	3.199	10.23	.1396	.01950	17.1	7.161	51.29
.6918	.4786	3.2	1.445	2.089	.3090	.09550	10.2	3.236	10.47	.1380	.01905	17.2	7.244	52.48
.6839	.4677	3.3	1.462	2.138	.3055	.09333	10.3	3.273	10.72	.1365	.01862	17.3	7.328	53.70
.6761	.4571	3.4	1.479	2.188	.3020	.09120	10.4	3.311	10.96	.1349	.01820	17.4	7.413	54.95
.6683	.4467	3.5	1.496	2.239	.2985	.08913	10.5	3.350	11.22	.1334	.01778	17.5	7.499	56.23
.6607	.4365	3.6	1.514	2.291	.2951	.08710	10.6	3.388	11.48	.1318	.01738	17.6	7.586	57.54
.6531	.4266	3.7	1.531	2.344	.2917	.08511	10.7	3.428	11.75	.1303	.01698	17.7	7.674	58.88
.6457	.4169	3.8	1.549	2.399	.2884	.08318	10.8	3.467	12.02	.1288	.01660	17.8	7.762	60.26
.6383	.4074	3.9	1.567	2.455	.2851	.08128	10.9	3.508	12.30	.1274	.01622	17.9	7.852	61.66
.6310	.3981	4.0	1.585	2.512	.2818	.07943	11.0	3.548	12.59	.1259	.01585	18.0	7.943	63.10
.6237	.3890	4.1	1.603	2.570	.2786	.07762	11.1	3.589	12.88	.1245	.01549	18.1	8.035	64.57
.6166	.3802	4.2	1.622	2.630	.2754	.07586	11.2	3.631	13.18	.1230	.01514	18.2	8.128	66.07
.6095	.3715	4.3	1.641	2.692	.2723	.07413	11.3	3.673	13.49	.1216	.01479	18.3	8.222	67.61
.6026	.3631	4.4	1.660	2.754	.2692	.07244	11.4	3.715	13.80	.1202	.01445	18.4	8.318	69.18
.5957	.3548	4.5	1.679	2.818	.2661	.07079	11.5	3.758	14.13	.1189	.01413	18.5	8.414	70.79
.5888	.3467	4.6	1.698	2.884	.2630	.06918	11.6	3.802	14.45	.1175	.01380	18.6	8.511	72.44
.5821	.3388	4.7	1.718	2.951	.2600	.06761	11.7	3.846	14.79	.1161	.01349	18.7	8.610	74.13
.5754	.3311	4.8	1.738	3.020	.2570	.06607	11.8	3.890	15.14	.1148	.01318	18.8	8.710	75.86
.5689	.3236	4.9	1.758	3.090	.2541	.06457	11.9	3.936	15.49	.1135	.01288	18.9	8.811	77.62
.5623	.3162	5.0	1.776	3.162	.2512	.06310	12.0	3.981	15.85	.1122	.01259	19.0	8.913	79.43
.5559	.3090	5.1	1.799	3.236	.2483	.06166	12.1	4.027	16.22	.1109	.01230	19.1	9.016	81.28
.5495	.3020	5.2	1.820	3.311	.2455	.06026	12.2	4.074	16.60	.1096	.01202	19.2	9.120	83.18
.5433	.2951	5.3	1.841	3.388	.2427	.05888	12.3	4.121	16.98	.1084	.01175	19.3	9.226	85.11
.5370	.2884	5.4	1.862	3.467	.2399	.05754	12.4	4.169	17.38	.1072	.01148	19.4	9.333	87.10
.5309	.2818	5.5	1.884	3.548	.2371	.05623	12.5	4.217	17.78	.1059	.01122	19.5	9.441	89.13
.5248	.2754	5.6	1.905	3.631	.2344	.05495	12.6	4.266	18.20	.1047	.01096	19.6	9.550	91.20
.5188	.2692	5.7	1.928	3.715	.2317	.05370	12.7	4.315	18.62	.1035	.01072	19.7	9.661	93.33
.5129	.2630	5.8	1.950	3.802	.2291	.05248	12.8	4.365	19.05	.1023	.01047	19.8	9.772	95.50
.5070	.2570	5.9	1.972	3.890	.2265	.05129	12.9	4.416	19.50	.1012	.01023	19.9	9.886	97.72
.5012	.2512	6.0	1.995	3.981	.2239	.05012	13.0	4.467	19.95	.1000	.01000	20.0	10.000	100.00
.4955	.2455	6.1	2.018	4.074	.2213	.04898	13.1	4.519	20.42					
.4898	.2399	6.2	2.042	4.169	.2188	.04786	13.2	4.571	20.89		10^{-3}	30		10^{3}
.4842	.2344	6.3	2.065	4.266	.2163	.04677	13.3	4.624	21.38	10^{-2}	10^{-4}	40	10^{2}	10^{4}
.4786	.2291	6.4	2.089	4.365	.2138	.04571	13.4	4.677	21.88		10^{-5}	50		10^{5}
.4732	.2239	6.5	2.113	4.467	.2113	.04467	13.5	4.732	22.39	10^{-3}	10^{-6}	60	10^{3}	10^{6}
.4677	.2188	6.6	2.138	4.571	.2089	.04365	13.6	4.786	22.91		10^{-7}	70		10^{7}
.4624	.2138	6.7	2.163	4.677	.2065	.04266	13.7	4.842	23.44	10^{-4}	10^{-8}	80	10^{4}	10^{8}
.4571	.2089	6.8	2.188	4.786	.2042	.04169	13.8	4.898	23.99		10^{-9}	90		10^{9}
.4519	.2042	6.9	2.213	4.898	.2018	.04074	13.9	4.955	24.55	10^{-5}	10^{-10}	100	10^{5}	10^{10}

NATIONAL BUREAU OF STANDARDS
CALIBRATION SERVICES AT RADIO FREQUENCIES

The following indicates in outline form the calibration services available from the Radio Standards Engineering Division, National Bureau of Standards, Boulder, Colorado. More detailed information is given in NBS Special Publication 250. Requests for this information, as well as on estimated costs of calibration, should be addressed to:

>Coordinator, Calibration Services
>Radio Standards Engineering Division
>National Bureau of Standards
>Boulder, Colorado 80302

Special services not listed in this tabulation often can be arranged upon request. Improvements in measurement accuracies, extension of frequency and magnitude ranges, and new types of calibrations are added to the services as they are developed. Requests to be on the mailing list of the Radio Standards Engineering Division for such information should be made to the above address.

COAXIAL STANDARDS, OR STANDARDS FITTED WITH COAXIAL CONNECTORS

Quantity	Interlaboratory Standards	Frequency Range	Magnitude Range	Estimated Limits of Uncertainty
ATTENUATION				
Insertion Loss	Fixed Attenuators	1, 10, 60, 100 MHz	0-80 dB	±0.05-5%
		30 MHz	0-100 dB	±0.05-5%
		0.10-12.4 GHz	0-100 dB	±0.05-5%
		12.4-18.0 GHz	0-80 dB	±0.05-5%
	Fixed Directional Couplers	1, 10, 60, 100 MHz	0-80 dB	±0.05-5%
		30 MHz	0-100 dB	±0.05-5%
		0.10-18.0 GHz	0-60 dB	±0.05-5%
Attenuation Difference	Variable Attenuators and Directional Couplers	1, 10, 60, 100 MHz	0-80 dB	±0.05-5%
		30 MHz	0-100 dB	±0.05-5%
		0.10-12.4 GHz	0-80 dB	±0.05-5%
		12.4-18.0 GHz	0-60 dB	±0.05-5%
	Waveguide Below-Cutoff Attenuators	1, 10, 60, 100 MHz	0-100 dB*	From 0.005 dB plus 0.05% of total dB to 0.05 dB plus 0.1% of total dB
		30 MHz	0-140 dB	
FIELD STRENGTH				
Antenna Factor	Loop Antennas	30 Hz-5 MHz	20-200 mV/meter	±0.25 dB
		5-30 MHz	20-200 mV/meter	±0.4 dB
	Dipole Antennas	30-1000 MHz	25-200 mV/meter	±1.0 dB
Voltage	Field Strength Meters Receivers as Two-Terminal RF Voltmeters	dc-400 MHz	0.001-10 mV	±3%
		400-1000 MHz	0.01-10 mV	±7%
		1-10 GHz	0.1-100 mV	depends on input VSWR of receiver
Attenuation Difference	Signal Attenuators	dc-400 MHz	0-80 dB	±0.1 dB + 0.3 of attenuation in dB
		400-1000 MHz	0-80 dB	±0.1 dB + 0.5 of attenuation in dB
		1-10 GHz	0-80 dB	depends on input VSWR of receiver
Linearity	Receivers	dc-400 MHz	0-60 dB (1-1000 μV)	±0.1 dB + 0.3 of attenuation in dB
		400-1000 MHz	0-60 dB (1-1000 μV)	±0.1 dB + 0.5 of attenuation in dB
		1-100 GHz	0-60 dB (1-1000 μV)	depends on input VSWR of receiver
FREQUENCY				
Frequency Measurement	Cavity Wavemeters	0.1-10 GHz		±10⁻⁴ to 10⁻⁶
Frequency Stability	Signal Sources	0-500 MHz		±10⁻¹¹
Power Spectral Analysis	Signal Sources	1, 2.5, 5, and 10 MHz		
IMMITTANCE				
Resistance	Two-Terminal Resistors (one-port)	30 kHz-2 MHz	0.1 Ω-1 MΩ	±0.1-4%
		2-5 MHz	50 Ω-1 MΩ	±0.1-6%
		5-100 MHz	20 Ω-50 kΩ	±0.1-10%
		5-250 MHz	20 Ω-20 kΩ	±0.2-10%

197

NATIONAL BUREAU OF STANDARDS
COAXIAL STANDARDS, OR STANDARDS FITTED WITH COAXIAL CONNECTORS

Quantity	Interlaboratory Standards	Frequency Range	Magnitude Range	Estimated Limits of Uncertainty
IMMITTANCE				
Inductance	Two-Terminal Inductors (one-port)	30 kHz-1MHz	$0.01\,\mu H$ to $\dfrac{10^{12}}{\omega^2}$ H	±0.1-20%
		1-5 MHz	$\dfrac{5 \times 10^8}{\omega^2}$ H to $\dfrac{10^{12}}{\omega^2}$ H	±0.1-0.5%
		5-250 MHz	$\dfrac{2 \times 10^{10}}{\omega^2}$ H to $\dfrac{10^{12}}{\omega^2}$ H	±0.1-5%
Capacitance	Two-Terminal Capacitors (one-port)	30 kHz-2 MHz	1 pF-0.1 μF	±0.1-5%
		2-5 MHz	1-1000 pF	±0.1-0.5%
		5-250 MHz	1-50 pF	±0.1-0.2%
	Three-Terminal Capacitors (two-port)	100-465 kHz, 1 MHz	0.01-1000 pF	±0.01-2%
Effective Q Effective Resonating Capacitance	Q Standards (banana plug connectors)	50 kHz-45 MHz	90-700 30-450 pF	+2-9 ±0.1-1 pF
Magnitude	Coaxial Loads, Mismatches, Lines, etc.	0.5-8 GHz	$1 \leqslant$ VSWR $\leqslant 100$ **$1 \leqslant$ VSWR < 4	±0.5-8% ±0.1-1%
Phase Angle			0-90° 0-90°	±0.5-1.5° ±0.1-1.0°
Magnitude of Reflection Coefficient	Coaxial Terminations	1-4 GHz	0-0.025 0.025-1	±0.00025 ±1% of $\lvert\Gamma\rvert$
NOISE				
Noise Temperature	Noise Generators	3 GHz	79 ± 4K, 373 ± 20K, 83-353K, 393-30,000K	+1%
PHASE SHIFT				
Phase Shift	Phase Shifters	30 MHz	0-360°	±0.1-0.5°
Phase Delay	Delay Lines	0.10-18 GHz	10^{-12} to 10^{-5} sec	0.5%
POWER, CW				
Calibration Factor	Bolometer Units	10, 30, 100, 200, 300, 400, 500, 700, 1000 MHz	1-10 mW	±1%
		1.3, 2, 2.2, 3, 3.5, 4 GHz		±1.5%
Effective Efficiency	Bolometer Units	10, 30, 100, 200, 300, 400, 500, 700, 1000 MHz	1-10 mW	±1%
		1.3, 2, 2.2, 3, 3.5, 4-17 GHz		±1.5% ±1-2%
Calibration Factor	Bolometer-Coupler Units	10, 30, 100, 200, 300, 400, 500 MHz	Coupling ratios 3-30 dB	±2%
		700, 1000 MHz, 1.3, 2, 2.2, 3, 3.5, 4 GHz	Coupling ratios 3-20 dB	±2%
Output Voltage/ Input Power	Calorimeters, RF Powermeters	10 MHz	0.001-200 W	±2%
		30 MHz	0.001-150 W	±2%
		100, 200, 300, 400, 500 MHz	0.001-100 W	±2%
		700, 1000 MHz	0.001-10 W	±2%
		1.3 GHz	0.001-50 W	±2%
		2, 2.2, 3, 3.5, 4 GHz	0.001-1 W	±2%
PULSE POWER				
Peak Pulse	Pulse Powermeters	300-500 MHz	0.001-2500 W (peak)†	±3%
		950-1100 MHz	0.001-3000 W (peak)†	±3%
		1100-1200 MHz	0.001-2500 W (peak)†	±3%
PULSE VOLTAGE				
Peak Pulse	Peak Pulse Voltmeters, Pulse Generators		5-100 V (peak)†† 100-1000 V (peak)†^	±1% ±1%
VOLTAGE				
Voltage, CW	RF Micropotentiometers, RF Millivoltmeters, RF Signal Sources	0.05-900 MHz	1-100,000 μV	±2-5%

NATIONAL BUREAU OF STANDARDS
COAXIAL STANDARDS, OR STANDARDS FITTED WITH COAXIAL CONNECTORS

Quantity	Interlaboratory Standards	Frequency Range	Magnitude Range	Estimated Limits of Uncertainty
VOLTAGE				
RF-DC Difference	Thermal Voltage Converters	0.03, 0.1, 0.3, 1, 3, 10, 30, 50, 100 MHz 300, 400, 500 MHz	0.1-200 V 0.1-15 V	±0.05-1% ±3%
	Attenuator- Thermoelement Voltmeters	10, 30, 100 MHz 300, 400, 500, 700, 900 MHz 1000 MHz	0.1-200 V 0.2-15 V 0.2-7 V	±1% ±3-7% ±7%

* Including initial insertion loss
** Employing precision 14mm coaxial connectors
† Trapezoidal pulses with:

Pulse duration range	2-10 μ sec
Pulse repetition rate range	100-1600 pps
Maximum duty factor	0.0033

†† Trapezoidal pulses with:

Amplitude range	5-100 V
Minimum rise and fall time	10 n sec (each)
Minimum peak duration	10 n sec
Pulse duration range	0.02-100 μ sec
Pulse repetition rate range	60-2 x 16^6 pps
Maximum duty factor	0.1
Amplitude range	100-1000 V
Minimum rise and fall time	30 n sec (each)
Minimum peak duration	30 n sec
Pulse duration range	0.06-5 μ sec
Pulse repetition rate range	60-1.66 x 10^5 pps
Maximum duty factor	0.01

WAVEGUIDE STANDARDS

Quantity	Interlaboratory Standards	Waveguide Size	Magnitude Range	Estimated Limits of Uncertainty
ATTENUATION				
Attenuation Difference	Variable Attenuators	WR 28 (26.5-40.0 GHz) WR 42 (18.0-26.5 GHz) WR 62 (12.4-18.0 GHz) WR 90 (8.20-12.4 GHz)	0 to 50 dB, extension to 70 dB in some waveguide sizes	Variable From ±0.005 dB to ±0.1 or 1%
Insertion Loss	Fixed Attenuators, (two-port)	WR 112 (7.05-10.0 GHz) WR 137 (5.85-8.20 GHz) WR 187 (3.95-5.85 GHz) WR 284 (2.60-3.95 GHz) WR 430 (1.70-2.60 GHz)		Fixed From ±0.1 dB or 1% to ±0.2 or 2%
FREQUENCY				
Frequency Measurement	Cavity Wavemeters	2.6 to 90 GHz		$\pm 10^{-4}$ to 10^{-6}
IMPEDANCE				
Reflection Coefficient Magnitude	Reflectors (mismatches), and Nonreflecting Waveguide Ports	WR 42 (18.0-26.5 GHz) WR 62 (12.4-18.0 GHz) WR 90 (8.20-12.4 GHz) WR 112 (7.05-10.0 GHz) WR 137 (5.85-8.20 GHz) WR 187 (3.95-5.85 GHz) WR 284 (2.60-3.95 GHz)	up to 0.2	For $\mid \Gamma \mid \geqslant 0.024$ $\pm (0.0002 + 0.002 \mid \Gamma \mid)$ For $\mid \Gamma \mid < 0.024$ $\pm (0.00013 + 0.0032 \mid \Gamma \mid)$
NOISE				
Effective Noise Temperature	Noise Sources	WR 62 (12.4-18.0 GHz) WR 90 (8.20-12.4 GHz) WR 284 (2.60-3.95 GHz)	1000-300,000 K (excess noise ratio range 3.8-30 dB)	±150-200 K
PHASE SHIFT				
Phase Shift Difference	Phase Shifters	WR 62 (12.4-18.0 GHz) WR 90 (8.2-12.4 GHz) WR 137 (5.85-8.20 GHz)	0-720°	±0.1 to 1°
POWER				
Effective Efficiency	Bolometer Units	WR 28 (26.5-40 GHz) WR 42 (18.0-26.5 GHz) WR 62 (12.4-18.0 GHz)	0.1-10 milliwatts	From ±0.5 to ±1.8%
Calibration Factor		WR 90 (8.20-12.4 GHz) WR 112 (7.05-10.0 GHz) WR 137 (5.85-8.20 GHz)		
Calibration Factor	Bolometer-Coupler Units	WR 187 (3.95-5.85 GHz) WR 284 (2.60-3.95 GHz)		

199

CHARACTERISTICS OF PRINCIPAL MICROWAVE RADIATION SAFETY STANDARDS EXPOSURE

Frequency Range:
USAS C95.1 - 10 MHz - 100 GHz
Military - all microwave frequencies - range not specified
USSR - 300 MHz to 30 GHz
Czech - 300 MHz to 300 GHz

Definition of Power Density:
Power densities referred to in standards is that average density measured in accessible regions (USASI, or military) or at actual exposure sites (USSR and Czech) in the absence of subject.

Averaging Time:
USAS C95.1 - 0.1 hour or 6 minutes
AF and Army - 0.01 hour or 36 seconds
Navy - 3 seconds
USSR - Not specified
Czech - Not specified, but the standard implies that an average density is calculated from an integrated dose. For example, for occupational situations the maximum permissible exposure is given by

$$\int_0^8 PdT < 200 \text{ microwatts/cm}^2 - \text{hours}$$

averaged over eight hours where P is power density and T is time in hours. The total exposure dose over five consecutive working days is summed and divided by 5 to obtain an average exposure dose for eight hours.

Dependence on Area of Exposure
No distinctions are generally made between partial and whole body exposure.*

Modification for Pulse or Other Modulation:
None except for reduction of exposure level by a factor of 2.5 in Czech standards.

Restrictions on Peak Power: None

Allowance for Environment:
None except for proposal by Mumford (6) to reduce the radiation exposure guide from 10 mw/cm² according to the formula $P_0(mw/cm^2) = 10 - (THI-70)$ for values of the temperature-humidity index (THI) in the range of 70 to 79 with $P_0 = 1$ mw/cm² for THI above 79.

Instrumentation:
Generally not well specified but far-field type probes such as small horns or open waveguides are specified with effective apertures $A_e = \dfrac{\lambda^2}{4\pi G}$ where G is the power gain. Response times are not well specified but are implied to be much greater than pulse durations and much smaller than duration of exposure - i.e., generally the order of seconds. Some use of true dosimetry - i.e., integrated absorbed energy is made in USSR and Czechoslovakia.

Under USSR standard exposure near 1 mw/cm² is permitted only with use of protective goggles for the eyes.

References
1. USA Standards Institute (now the American National Standards Institute, New York): "Safety Level of Electromagnetic Radiation with Respect to Personnel," USAS C95.1 - 1966.
2. "Control of Hazards to Health from Microwave Radiation," Army Technical Bulletin TB Med .270, Air Force Manual AFM 161-7, Dec. 1965.
3. "Technical Manual for RF Radiation Hazards," Dept. of Navy, Naval Ships System Commands, NAVSHIPS 0900-005-8000, July 1966.
4. "Temporary Safety Regulations for Personnel in the Presence of Microwave Generators," USSR Ministry of Hygiene, Publication No. 273-58, Nov. 26, 1958.
5. "Supplement to the Information Bulletin for the Discipline of Industrial Hygiene and Occupational Diseases and for Radiation Hygiene," Prague, June 1968.
6. W. W. Mumford, "Heat Stress due to RF Radiation," Proc. IEEE, 57, pp. 171-178, Feb. 1969.

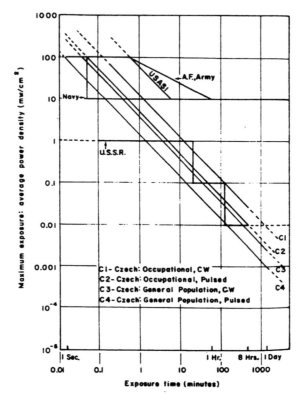

C1- Czech: Occupational, CW
C2- Czech: Occupational, Pulsed
C3- Czech: General Population, CW
C4- Czech: General Population, Pulsed

Courtesy of Dr. John Osepchuk

CHARACTERISTICS OF PRINCIPAL MICROWAVE OVEN LEAKAGE PERFORMANCE STANDARDS

HEW Standard:
>   Frequency Range: 890-6000 MHz
>   Maximum Power Density*: 5 mw/cm^2 in fields,
>   1 mw/cm^2 at factory as measured at any point
>   2.0 inches from external surface of oven.
>   Instrumentation: Effective aperture of measurement
>   probe: less than 25 cm^2.
>   Averaging time: 3 sec.

AHAM Standard:
>   Frequency Range: 915 (890-940) MHz and 2450
>   (2400-2500) MHz.
>   Maximum Power Density*: 10 mw/cm^2 as measured
>   at any point 2.0 inches from external surface of
>   oven.
>   Instrumentation: Effective aperture of measurement
>   probe: Two inch maximum as defined by a
>   proximity effect test resulting in a 20 percent
>   change in probe reading when a second probe
>   is present.

*Power density is an effective power density related to measured E field through far-field plane wave formulas.

References
1. Proposed Performance Standard, "Microwave Ovens," Control of Electronic Product Radiation, PHS, HEW, Federal Register, Vol. 35, No. 100, pages 7901-7902, Friday, May 22, 1970.
2. Association of Home Appliance Manufacturers, Chicago, "Safety and Measurement Standard: Microwave Ovens," Feb. 1970, in process of administrative approval.

Typical Spatial Distribution of Leakage Near Microwave Ovens

Leakage fields generally exhibit a 1/r dependence of power density close to oven (<6.0") and a 1/r^2 dependence further away (>1 foot). Actual distributions will show minor fluctuations (e.g., up to 3 dB) around smooth 1/r^2 extrapolations.

Estimated upper limits shown in figure approximate the maximum for field radiation assuming radiation along entire door perimeter at the maximum permitted level into a hemisphere and neglecting door screen leakage. (This is generally negligible but would increase far field levels by less than 3 dB even if entire screen leaked at maximum level).

The estimated lower limit approximtaes the distribution resulting from one localized source. Typical measured distributions fall between these estimated limits; when measured along a line normal to point of maximum leakage.

A reasonable general estimate is a simple 1/r^2 extrapolation from the measured power density at 2.0 inches - points A, B or C. For example, such an extrapolation from point B happens to coincide with far-field upper limit associated with point C.

These estimated distributions are for the case of an oven leaking at its maximum permitted level as specified in the performance standard.

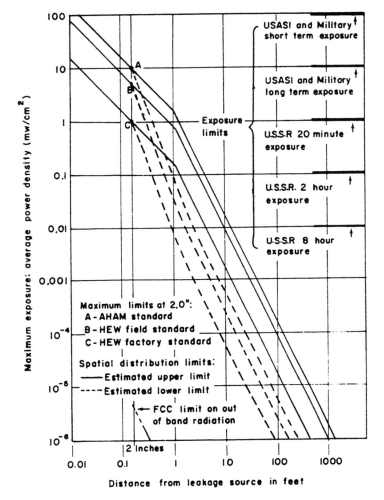

Courtesy of Dr. John Osepchuk

VELOCITY OF PROPAGATION FOR ELLIPTICAL WAVEGUIDE

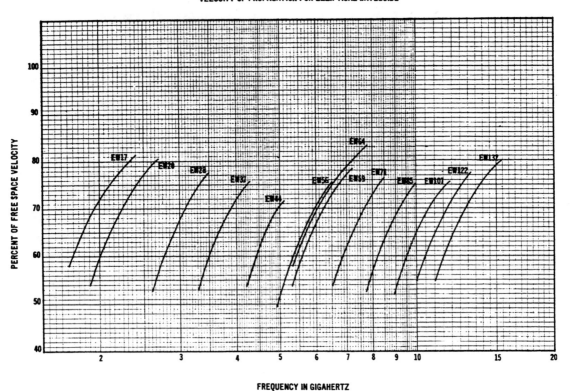

FREQUENCY IN GIGAHERTZ

VELOCITY OF PROPAGATION FOR RECTANGULAR WAVEGUIDE

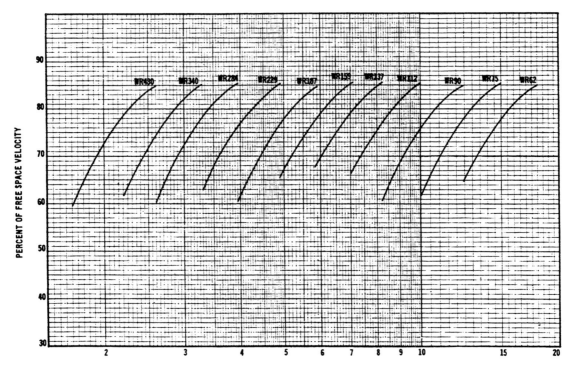

FREQUENCY IN GIGAHERTZ

Courtesy of Dr. John Osepchuk

Microwave Engineers'

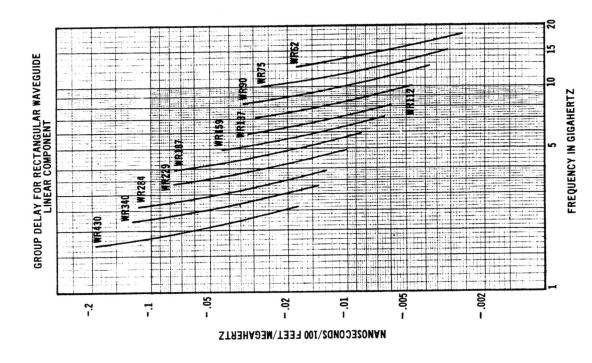

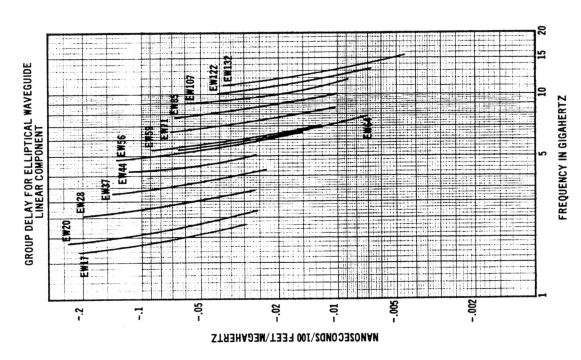

Courtesy of Dr. John Osepchuk

203

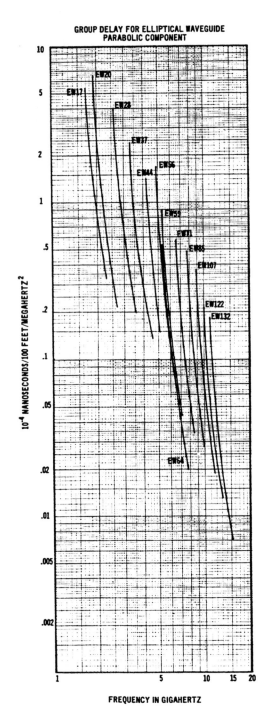

GROUP DELAY FOR ELLIPTICAL WAVEGUIDE
PARABOLIC COMPONENT

FREQUENCY IN GIGAHERTZ

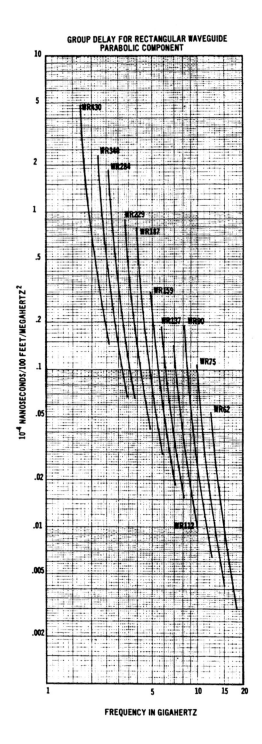

GROUP DELAY FOR RECTANGULAR WAVEGUIDE
PARABOLIC COMPONENT

FREQUENCY IN GIGAHERTZ

Courtesy of Dr. John Osepchuk

INTERMODULATION DISTORTION IN FDMFM
MICROWAVE RADIO RELAY SYSTEMS DUE TO FEEDER MISMATCH

It is now well known[1] that the presence of any non linear phase versus frequency characteristic or any multipath effect in a Frequency Division Multiplexed Frequency Modulated (FDMFM) microwave communication link distorts the modulation spectrum. One such multipath effect is the echo signal that occurs due to an imperfect match at the antenna and equipment ends of the R.F. feeder line.[2],[3],[4],[5],[6] This modulation spectrum distortion manifests itself, for example, by causing one telephone channel to interfere with another. Thus, even though an empty channel is transmitted, after reception it will contain many contributions from all the remaining active channels. This unwanted signal is referred to as intermodulation distortion. The contribution made to the total intermodulation distortion due to a feeder line having only end mismatches, as depicted in Fig. 1, comes about because the original signal leaving the antenna, e(t), has added to it the echo signal $r\ e(t-\tau)$ due to the end reflections, where:

(1) $r = \Gamma_1\ \Gamma_2 e^{-2\alpha L}$ = Echo Amplitude

(2) Γ_1 = Reflection Coefficient Looking Into Antenna

(3) Γ_2 = Reflection Coefficient Looking Into Equipment

(4) α = One Way Attenuation of Feeder Cable, nepers/foot

(5)   L = Length of Feeder Cable, feet.

(6) $\tau = \dfrac{2L}{v_g}$ Round Trip Transit Time in Feeder Cable Echo Delay Time, secs.

(7) v_g = Group Velocity in Cable, feet/sec.

Under these conditions, the maximum signal distortion, D, that this produces in the top channel lies below the echo signal, r^2S, in the manner determined by Medhurst[5], as shown in Fig. 2. Here:

(8) $\Theta = p_m \tau$ radians

(9) p_m = Maximum Baseband Modulating Frequency, radians/sec.

(10) Δ = Total R.M.S. Frequency Deviation For All Channels, radians/sec.

(11) $A = \dfrac{\Delta}{p_m}$ Dimensionless Parameter (Ratio of Δ to p_m)

The curves in Fig. 2 (which are Fig. 1 of Medhurst[5]) show that for a fixed microwave channel (A and p_m fixed) that as the cable length increases, the distortion increases monotonically up to a certain cable length after which it flattens out somewhat in an oscillatory manner. The effect of cable attenuation is taken into account by realizing that the echo level is r^2S (i.e., below the signal level, S, by r^2), and the distortion, D, is down from the echo level, by the amount shown in Fig. 2, which are plots of $10\ log_{10}(\frac{D}{r^2S})$.

As a specific realistic example, the level of maximum distortion in the top channel of an 1800 channel system with CCIR values of p_m = 5.0278·10^7 rads./sec. (corresponding to a maximum base band frequency of 8.002 MHz), Δ = 6.635·10^6 rads./sec. (corresponding to a total RMS frequency deviation of 1.056 MHz), and A = 0.132, is plotted in Fig. 3 (which is based on Andrew Corporation's recent independent rederivation and calculations of equation (5) of Medhurst[5], since his Fig. 1 does not specifically contain this example). (For this example, the Θ abscissa multiplied by 10 corresponds approximately to the feeder length in feet if the group velocity in the cable is approximately the speed of light in vacuum. Then the abscissa range in Θ of $0.1 \le \Theta \le 20.0$ corresponds to a cable of approximately 1 to 200 feet). Thus, for example, for a cable run of 10 feet, the maximum distortion is down approximately 25 dB from the echo level, whereas for a run of 70 feet it is only down approximately 5 dB. If the end mismatches are Γ_1 = 0.025, Γ_2 = 0.04, and the cable has negligible attenuation, $e^{-2\alpha L}$ = 1, then r^2 = 10^{-6}, i.e., the echo is down 60 dB from the signal level S. For the CCIR signal power per channel of S = -15 dBm0 (unweighted) the distortion is then -100 dBm0 or 0.10 pWO for L = 10 feet, and -80 dBm0 for 10 pWO for L = 70 feet (where pWO is picowatts unweighted). These results are for no pre-emphasis. It is noted that the above curves give the maximum distortion which can occur, the average distortion will be less. In general, an echo distortion power of 10 pWO represents a very good system.

1.   D. H. Hamsher, "Communication System Engineering Handbook," McGraw Hill Book Co., N.Y., 1967, Chapter 16, pp.23-30.

2.   L. Lewin, "Phase Distortion in Feeders," Wireless Engineer, V.27 pp.143-145, May 1950.

3.   W. J. Albersheim and J. P. Schafer, "Echo Distortion in the F.M. Transmission of Frequency-Division-Multiplex," Proc. IRE, Vol.40 pp.316-328, March, 1952.

4.   W. R. Bennett, H. E. Curtiss, and S. O. Rice, "Interchannel Interference in FM and PM Systems Under Noise Loading Conditions," BSTJ, Vol.34, pp.601-636, May, 1955.

5.   R. G. Medhurst, "Echo Distortion in Frequency Modulation," Electronic and Radio Engineer, July, 1959, pp.253-259.

6.   I. Noodt, "Propagation Computer Saves Time," Microwaves, Vol.7, No. 8, Aug. 1968, pp.49-52.

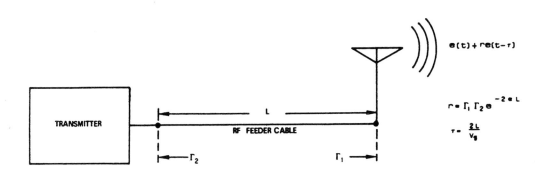

$e(t) + r e(t-\tau)$

$r = \Gamma_1 \Gamma_2 e^{-2\alpha L}$

$\tau = \dfrac{2L}{v_g}$

TRANSMITTER

L

RF FEEDER CABLE

Γ_2 Γ_1

FIGURE 1 - ECHO ESTABLISHMENT IN AN ANTENNA FEEDER

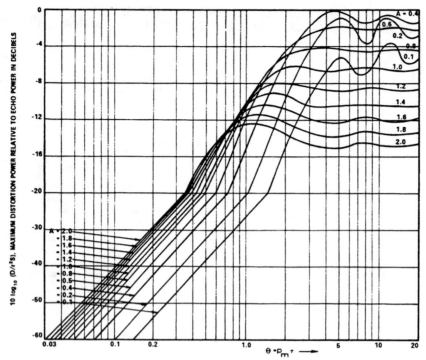

FIGURE 2 - MAXIMUM ECHO DISTORTION POWER IN TOP CHANNEL RELATIVE TO ECHO AMPLITUDE

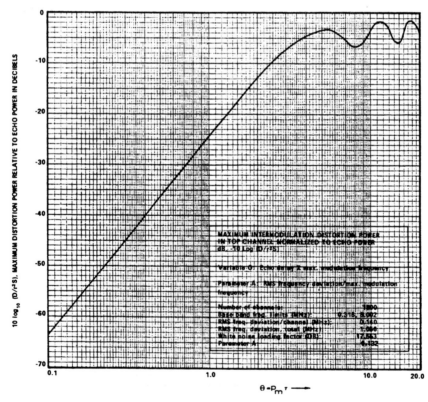

FIGURE 3 - MAXIMUM ECHO DISTORTION IN AN 1800 CHANNEL SYSTEM

Courtesy of Andrew Corporation, Orland Park, Illinois

MICROWAVE PASSIVE REPEATER NOMOGRAPH

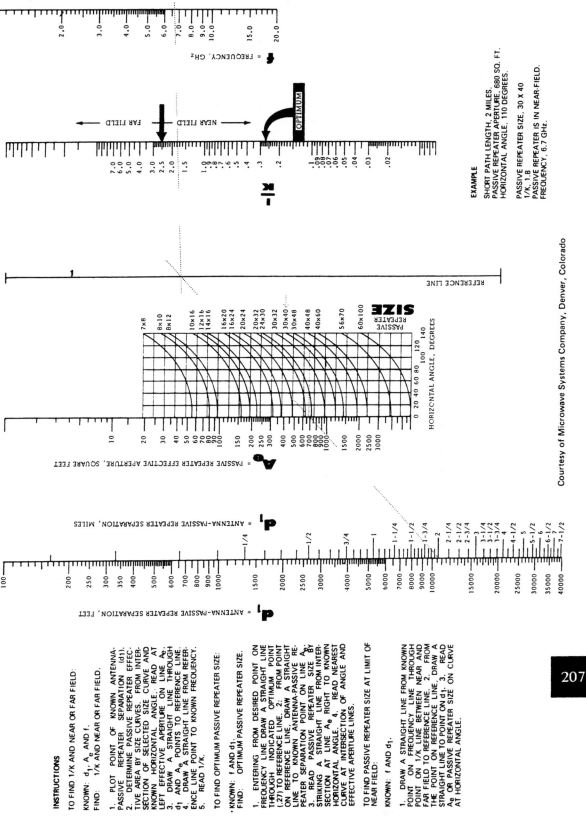

f = FREQUENCY, GHz

FAR FIELD ← | → NEAR FIELD

OPTIMUM

$\frac{1}{K}$

REFERENCE LINE

SIZE

PASSIVE REPEATER

HORIZONTAL ANGLE, DEGREES

A_e = PASSIVE REPEATER EFFECTIVE APERTURE, SQUARE FEET

d_1 = ANTENNA-PASSIVE REPEATER SEPARATION, MILES

d_1 = ANTENNA-PASSIVE REPEATER SEPARATION, FEET

INSTRUCTIONS

TO FIND 1/K AND NEAR OR FAR FIELD:

KNOWN: d_1, A_e AND f.
FIND: 1/K AND NEAR OR FAR FIELD.

1. PLOT POINT OF KNOWN ANTENNA-PASSIVE REPEATER SEPARATION (d_1). 2. DETERMINE PASSIVE REPEATER EFFECTIVE AREA BY SIZE CURVES. FROM INTERSECTION OF SELECTED SIZE CURVE AND KNOWN HORIZONTAL ANGLE, READ AT LEFT EFFECTIVE APERTURE ON LINE A_e. 3. DRAW A STRAIGHT LINE THROUGH d_1 AND A_e POINTS TO REFERENCE LINE. 4. DRAW A STRAIGHT LINE FROM REFERENCE LINE POINT TO KNOWN FREQUENCY. 5. READ 1/K.

TO FIND OPTIMUM PASSIVE REPEATER SIZE:

KNOWN: f AND d_1.
FIND: OPTIMUM PASSIVE REPEATER SIZE.

1. ENTERING FROM DESIRED POINT ON FREQUENCY LINE DRAW A STRAIGHT LINE THROUGH INDICATED OPTIMUM POINT (.27) TO REFERENCE LINE. 2. FROM POINT ON REFERENCE LINE, DRAW A STRAIGHT LINE TO KNOWN ANTENNA-PASSIVE REPEATER SEPARATION POINT ON LINE d_1. 3. READ PASSIVE REPEATER SIZE BY STRIKING A STRAIGHT LINE FROM INTERSECTION AT LINE A_e RIGHT TO KNOWN HORIZONTAL ANGLE. 4. READ NEAREST CURVE AT INTERSECTION OF ANGLE AND EFFECTIVE APERTURE LINES.

TO FIND PASSIVE REPEATER SIZE AT LIMIT OF NEAR FIELD:

KNOWN: f AND d_1.

1. DRAW A STRAIGHT LINE FROM KNOWN POINT ON FREQUENCY LINE THROUGH POINT ON 1/K LINE BETWEEN NEAR AND FAR FIELD TO REFERENCE LINE. 2. FROM THE POINT ON REFERENCE LINE, DRAW A STRAIGHT LINE TO POINT ON d_1. 3. READ A_e OR PASSIVE REPEATER SIZE ON CURVE AT HORIZONTAL ANGLE.

EXAMPLE

SHORT PATH LENGTH, 2 MILES.
PASSIVE REPEATER APERTURE, 680 SQ. FT.
HORIZONTAL ANGLE, 110 DEGREES.

PASSIVE REPEATER SIZE, 30 X 40
1/K, 1.8
PASSIVE REPEATER IS IN NEAR-FIELD.
FREQUENCY, 6.7 GHz.

Courtesy of Microwave Systems Company, Denver, Colorado

MICROWAVE ACOUSTICS
M.F. LEWIS

DESCRIPTION OF DEVICE
Typical Two-port Microwave Ultrasonic Delay Line

The delay line consists of a single crystal delay medium with optically flat and parallel end faces on to which electrodes and transducers are deposited as shown. It is necessary to prepare the delay lines to the highest optical tolerances because of the very short acoustic wavelengths at microwave frequencies (comparable to optical wavelengths), see Fig. 2. Acoustic bonds cannot be used at microwave frequencies because they are too lossy. Similarly the delay medium must be a single crystal, all other materials being too lossy. Typical material losses are listed in Table I.

BANDWIDTH CONSIDERATIONS

Each half of the device (input or output) has an acoustic and an electrical bandwidth. The acoustic bandwidth depends primarily on the acoustic impedances, Z, (Z = ρc where ρ = material density and c = velocity of the sound wave under consideration) and thicknesses of the various components (VIZ electrodes, piezoelectric transducer and delay medium). The acoustic bandwidth is calculated by considering the two faces of the transducer to be sources of

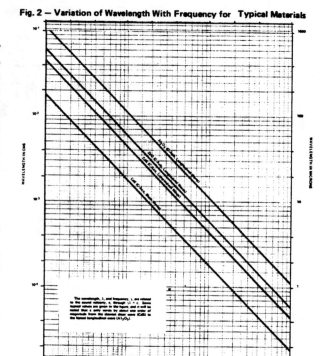

Fig. 2 — Variation of Wavelength With Frequency for Typical Materials

sound (180° out of phase) and each component to be a section of (acoustic) transmission line with appropriate impedance Z. Typical values of ρ, c and Z are included in the table.

Each transducer has an electrical equivalent circuit which consists primarily of its static capacitance in parallel with a radiation impedance. This latter arises from the piezoelectric coupling together with acoustic loading of the transducer by the rest of the delay line. In general some sort of electrical tuning will be employed, and the main design problem is to optimize the trade-off between bandwidth and insertion loss. Typical state-of-the-art performance is given by Olson (Fig. 3). The best transducers currently available employ ZnO.

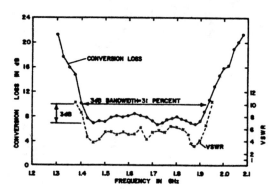

Fig. 3 — State-of-the-Art Delay Line Performance

Courtesy of F. A. Olson

Material	Acoustic Mode	Velocity (cm/sec)	Z_0 K$_g$/sec m²	Attenuation (dB/μsec at 1.5 GHz)	Approximate Max Length and Delay (cm and μsec)	Comments
Fused Silica	Longitudinal	5.96 x 10⁵	13.1 x 10⁶	22	*	*100's μsec in polygonal crystals at low frequencies.
Fused Silica	Shear	3.76 x 10⁵	8.3 x 10⁶	16	*	
X-cut Quartz	Longitudinal	5.75 x 10⁵	15.2 x 10⁶	5.6	12, 20	Excellent quality.
Z-cut Quartz	Longitudinal	6.32 x 10⁵	16.7 x 10⁶	3.4	12, 20	Free from spurious modes.
AC-cut Quartz	Shear	3.32 x 10⁵	8.8 x 10⁶	4.0	12, 35	Long maximum delay due to slow velocity.
BC-cut Quartz	Shear	5.06 x 10⁵	13.4 x 10⁶	1.65	12, 25	
Sapphire	Longitudinal	11.1 x 10⁵	43.6 x 10⁶	0.75	6, 6	Useful at high frequencies.
Rutile	Shear	5.4 x 10⁵	23.0 x 10⁶	0.97	3, 6	Useful at high frequencies.
LiNbO₃	Longitudinal	7.4 x 10⁵	34.8 x 10⁶	0.80	3, 4	Uncertain quality.
YIG	Shear	3.87 x 10⁵	20.0 x 10⁶	0.6	2, 5	Size limited.
YAG	Shear	5.03 x 10⁵	22.9 x 10⁶	0.36	5, 10	Applicable for long delay at high frequencies.
YAG	Longitudinal	8.56 x 10⁵	38.9 x 10⁶	0.6	5, 6	

TABLE I — Properties of Microwave Acoustic Materials

DELAY TIME CONSIDERATIONS

Since acoustic propagation is non-dispersive the delay time, T, for a pathlength, I, is simply given by $T = I/c$. Typical values of c are given in the table. The delay time per cm thus varies from about 1 to 5 μ sec.

INSERTION LOSS CONSIDERATIONS

At the lower microwave frequencies (~ 1 GHz) the transducer loss dominates for delays below about 10 μ sec, and two-way insertion losses are commonly of order 20 dB for moderate bandwidth. (~20%).

At higher frequencies (~ 9 GHz) material propagation losses predominate for $T \gtrsim 1 \mu$ sec. The attenuation is caused by interactions with thermal vibrations of the lattice and at room temperature is given by

$$\alpha(dB/cm) = \frac{5 \delta^2 C T_0 \omega^2 \tau}{\rho c^3}$$

where C = specific heat/unit volume, $T_0 = 300°$ K, $\omega =$ ultrasonic radian frequency, δ^2 is a material constant of order unity, and τ is a relaxation time approximately given by

$$\tau = 3K/C\bar{c}^2$$

where K = thermal conductivity and $\bar{c}$ = average sound wave velocity for the material.

Note that α varies as ω^2, as shown in Fig. 4. For a typical material at 9 GHz, $\delta^2 = 1$, $c = \bar{c} = 5 \times 10^5$ cm/sec, $\rho = 3$ gm/cm^3, $K = 10^6$ erg/cm sec $°$K giving α 150 dB/cm, as commonly found. Each quantity in the formula for α can vary by a factor of 2-3 between different materials, and the resulting range of attenuations covers 2-3 orders of magnitude (see table). This attenuation can be removed by cooling the crystal to freeze out the thermal vibrations, see Fig. 5.

Generally, materials with high Debye temperatures (Θ_D) have (i) lower room temperature losses and (ii) the attenuation drops (say to half its room temperature value) at higher temperatures. This is illustrated in Figs. 5 and 6

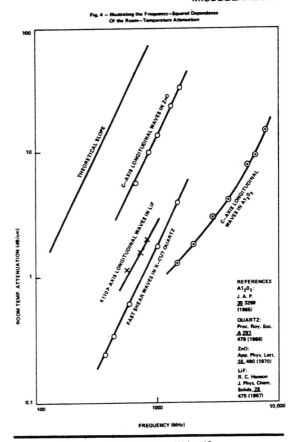

Fig. 4 – Illustrating the Frequency–Squared Dependence Of the Room–Temperature Attenuation

REFERENCES

A1$_2$0$_3$:
J. A. P.
36 3209
(1965)

QUARTZ:
Proc. Roy. Soc.
A 293
479 (1966)

ZnO:
App. Phys. Lett.
16, 480 (1970)

LiF:
R. C. Hanson
J. Phys. Chem.
Solids, 28
475 (1967)

Note: This graph is an extended version of K. Dransfeld, Journal De Physique, Colloque C1, Supplement au No. 2, Tome 28, Feb 1967 page C1-157 Fig. 5 for cubic materials propagation is usually along <100>. For others, propagation is along the C–Axis, except for SiO$_2$ and A1$_2$0$_3$ (propagation on A–Axis) when the lower shear wave attenuation is plotted.

Fig. 6 – Variation of Acoustic Attenuation with Debye Temperature (Θ_D)

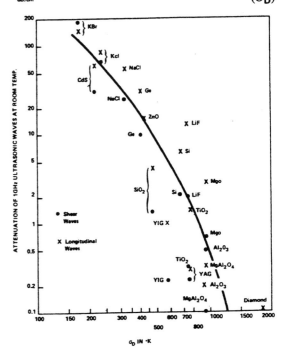

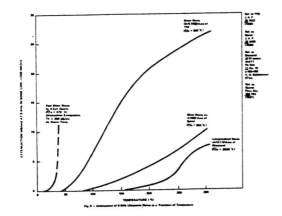

Fig. 5 – Attenuation of 9 GHz Ultrasonic Waves as a Function of Temperature